Insect predator–prey dynamics

Much of our understanding about insect predator–prey dynamics derives
from studies on insect parasitoids. But do true predators such as ladybird
beetles really operate in a similar way and how does this affect their use
in biological control? The extensive literature on ladybirds as biocontrol
agents shows that their size and rate of development is very dependent
on the nature of their prey. This volume explores the basic biology of
ladybirds, their association with their prey and its effect on development
rate and body size. Optimal foraging theory, field observations and
laboratory experiments are used to illustrate how ladybird larvae
maximize their rate of energy intake, and ladybird adults their fitness.
The interdependence of these life-history parameters is then used to
develop a simple predator–prey model, which with an analysis of the
literature highlights the specific attributes of potentially successful
biocontrol agents for all those interested in predator–prey dynamics.

A.F.G. DIXON is an Emeritus Professor in the School of Biological
Sciences at the University of East Anglia. He has written over 200 papers
on aphids and their natural enemies in scientific journals, and has
written or edited nine other books. In 1992, he was awarded the Gregor
Mendel Gold Medal by the Czech Academy of Science.

Insect predator–prey dynamics

Ladybird beetles and biological control

A.F.G. DIXON

University of East Anglia

CAMBRIDGE UNIVERSITY PRESS
Cambridge, New York, Melbourne, Madrid, Cape Town, Singapore, São Paulo

Cambridge University Press
The Edinburgh Building, Cambridge CB2 2RU, UK

Published in the United States of America by Cambridge University Press, New York

www.cambridge.org
Information on this title: www.cambridge.org/9780521622035

First published 2000
This digitally printed first paperback version 2005

A catalogue record for this publication is available from the British Library

Library of Congress Cataloguing in Publication data
Dixon, A. F. G. (Anthony Frederick George)
  Insect predator–prey dynamics : ladybird beetles and biological control /
  A. F. G. Dixon.
     p.   cm.
  Includes bibliographical references (p.   ).
  ISBN 0 521 62203 4
  1. Ladybugs. 2. Predation (Biology) 3. Insect pests–Biological control. I. Title.
QL596.C65 D58 2000
595.76′9   99-045440

ISBN-13 978-0-521-62203-5 hardback
ISBN-10 0-521-62203-4 hardback

ISBN-13 978-0-521-01770-1 paperback
ISBN-10 0-521-01770-X paperback

Contents

	Preface	ix
1	**Introduction**	1
2	**Basic biology and structure**	6
	Introduction	6
	Life cycle	8
	Morphology	10
	Mouthparts	12
	Alimentary canal	14
	Legs	14
	Development	15
	Survival	17
	Reproduction	18
	Fecundity and longevity	18
	Effect of food supply on egg and cluster size	21
	Interspecific relationships	23
	Overwintering	27
	Defence	29
3	**Body size**	36
	Introduction	36
	Intraspecific plasticity in size	36
	Sex and size	42
	Sexual size dimorphism	44
	Protandry	46
	Gonadal constraint	50
	Fecundity advantage	52
	Time and energy constraint	52
	Body size distribution	55
	Theory	55
	Empirical data	57

	Prey of ladybirds	58
	What determines the shape of the size diversity curves?	64
4	**Slow–fast continuum in life history parameters**	67
	Introduction	67
	Speed of movement	68
	Developmental time and metabolic rate	69
	Fecundity and longevity	74
	Trade-off	76
5	**Foraging behaviour**	82
	Introduction	82
	Functional response	82
	Prey recognition	83
	Relative risk as a determinant of diet breadth	84
	Prey specificity	88
	Switching	93
	Adult foraging behaviour	95
	Location of prey	97
	Patch quality	102
	Egg distribution	108
	Larval foraging behaviour	109
	Location of prey	109
	Survival	124
	Ladybird abundance	126
6	**Cannibalism**	130
	Introduction	130
	Theory	130
	Cannibalism by adults	133
	Cannibalism by larvae	134
	Fitness	135
	Currency	137
	Model	138
	Empirical data	140
	Avoidance of cannibalism	142
	Eggs	142
	Pupae	146
	Cannibalism as a means of harvesting prey – the icebox hypothesis	147
7	**Theory of predator–prey interactions**	151
	Introduction	151

	Parasitoids	151
	Predators	151
	Theory	152
	Nicholson & Bailey	155
	Optimal foraging approach	160
	Minimalism – generation time ratios	165
8	**Intraguild predation**	**173**
	Introduction	173
	Guild structure	173
	Predator–predator interactions	175
	Additive/non-additive effects	175
	Predator facilitation	179
	Top predators	179
	Cost of intraguild predation	183
9	**Biological control**	**190**
	Introduction	190
	Conflicts of interest	192
	Theory	193
	Biological control and conservation	196
	Biological control agents	197
	Ladybirds and biological control	199
	Attributes of successful biological control agents	200
	Augmentative biological control	211
	Cultural control	212
	Integrated pest management	214
	Is biological control evolutionarily stable?	216
10	**Epilogue**	**218**
11	**References**	**220**
	Taxonomic index	**253**
	Subject index	**255**

Preface

Ladybird beetles are familiar and popular insects and therefore need no introducing. The objective of this book is to give university students and research workers a better understanding of predation by insects than is to be found in most current ecological texts.

As stated in the Introduction the foundations of this book were built upon the enthusiasm for and understanding of ladybirds of my colleagues and students. In addition Jean-Louis Hemptinne kindly read and commented on the whole manuscript, and Pavel Kindlmann, on Chapter 7. Other people too have helped in the preparation of the book. I am especially grateful to CSIRO Entomology and Veronica Brancatini for supplying, with permission to use, the photograph of *Rodolia cardinalis* that is on the back cover, and to CAB International and Roger Booth for permission to reproduce the habitus drawing of *Hyperaspis pantherina* in Chapter 9. I also wish to express special thanks to Diane Alden for preparing the figures and to Karen Harris for typing some of the manuscript.

I dedicate this book to June.

Tony Dixon

1

Introduction

Why ladybirds function thus and not otherwise.

Ivo Hodek, 1997

Facts without theory is chaos, theory without facts is fantasy.

The first and most successful case of classical biological control was the introduction in 1888 of the Australian ladybird, *Rodolia cardinalis*, into California, for the suppression of cottony-cushion scale, *Icerya purchasi*. The threat to citrus production posed by this scale insect was successfully averted (Doutt, 1964) and over one hundred years on this beetle is still effective in keeping the numbers of the scale well below the economic threshold. This outstanding success resulted in the widespread and haphazard introduction of natural enemies, especially ladybirds, which has been referred to as the 'ladybird fantasy' period (Lounsbury, 1940). This was in part also possibly fuelled by the way ladybirds had been perceived for centuries. They are often seen as harbingers of good weather or fortune. In Czech they are called *sluníčka* – small suns, and in Japanese *tento mushi* – sun-loving insects. The Vikings dedicated the seven-spot ladybird to the wife of Odin, Frigg, the goddess of domestic conjugal love, and called it *Friggahönna*. Generally for Norsemen ladybirds were believed to predict the harvest. If they had more than seven spots, bread would be dear, if seven or fewer the harvest would be abundant and prices low. Interestingly, many of the species with large numbers of spots are fungal feeders and their presence in noticeable numbers would indicate a high incidence of fungal diseases, and if associated with cereals a poor grain yield. After the rise of Christianity Frigg was replaced by the Holy Virgin and so the present names originated. The prevalence of holy attributes in their common names in all European languages possibly indicates a widely held belief that they are harbingers of good tidings.

When I first studied ladybirds forty years ago I also was impressed by the success of *Rodolia* in dramatically reducing the abundance of *Icerya*, especially as this contrasted very markedly with my results, which indicated that the ten-spot ladybird was a very ineffective predator of aphids (Dixon, 1959), and with the many unsuccessful biological control programmes against aphids involving the use of ladybirds. This stimulated the question: why are some ladybirds successful biocontrol agents and others not? An increasing interest in aphids distracted me from thinking more deeply about this problem. However, I have been fortunate in that three students – Michael Carter, Nick Mills and Lesley Stewart – and seven colleagues – Basant Agarwala, Ted Evans, Ivo Hodek, Zdenek Růžička, Hironori Yasuda, and above all Jean-Louis Hemptinne and Pavel Kindlmann – have kept my interest in ladybirds alive and made me think more deeply about the paradox posed by the marked differences in the ability of ladybirds to suppress the abundance of their prey. Ted, in convincing me that I should pay more attention to the oviposition behaviour of adults, played a major role in initiating the conceptual approach to ladybird foraging strategies adopted here.

The books on ladybirds by Hodek & Honěk (1996), Klausnitzer & Klausnitzer (1997) and Majerus (1994) give excellent general accounts and coverage of the extensive literature. However, in attempting a complete coverage they have not the freedom to explore any one topic in depth, although the particular interests of the authors are reflected in the numbers of pages devoted to particular topics. There have been three reviews of the literature on predaceous ladybirds, the first by Hagen (1962), the second by Hodek (1967) and the third by Obrycki & Kring (1998). Both Hagen and Hodek comment on the effectiveness of ladybirds in classical biological control and agree that aphidophagous species are less effective than coccidophagous species. Hagen is generally more positive about their role with the qualification that aphidophagous species cannot be relied on for control at certain times. Hodek attributes the poor success of aphidophagous species to their slow rate of development compared with that of their prey and to their arriving too late. However, when acting in combination with other natural enemies of aphids they are believed to be effective. Obrycki & Kring, although noting that coccinellids have been widely used in biological control for over a century, did not account for the successes and failures, but concentrated on technology and practice rather than the pattern and process of ladybird–prey interactions.

Although the results of studies on ladybirds are frequently used by theoreticians to illustrate the types of foraging behaviour shown by insect predators, the dynamics of ladybird–prey interactions have rarely been rigorously analysed. Crawley (1992), however, in his book *Natural Enemies* uses a ladybird–aphid study to illustrate his chapter on population dynamics of natural enemies and their prey. The study cited was that of Frazer & Gilbert (1976), which analyses the interaction between *Coccinella trifasciata* and the aphid *Acyrthosiphon pisum* on alfalfa in the field. Crawley follows the authors of this study in attributing the ineffectiveness of the ladybird to it not having enough time to find sufficient prey so that prey survival rate never fell to zero. He also follows the authors in championing the suggestion that the most important aspect of insect predator–prey dynamics is the difference in the lower temperature thresholds of the predator and prey. As the ladybird's lower temperature threshold is substantially higher than the aphid's the ladybird is unlikely to have a significant impact on the aphid's abundance, because the ladybird always arrives too late to prevent aphid population build-up. As the thermal thresholds of insects are likely to be subject to natural selection (cf. Campbell *et al.*, 1974) it is relevant to ask why the lower thermal threshold of ladybirds should differ from that of their prey. The suspicion is that the difference is adaptive rather than maladaptive as implied above. Similarly, Gutierrez (1996) uses data collected for aphidophagous ladybirds to build a general model of predator–prey interactions the predictions of which are compared with the dynamics of ladybird–aphid and ladybird–coccid interactions observed in nature. Although he pursues a very commendable aim in attempting to draw out the commonalities in such interactions and generalizing, he like others (e.g. Kareiva & Odell, 1987; Skirvin *et al.*, 1997) gives little consideration to whether the behaviours ascribed to adult predators and incorporated in their models, in particular the aggregative and reproductive responses, are realistic and above all adaptive. This is likely to result in erroneous conclusions.

Although predators clearly differ from parasitoids in their individual killing potential and specificity, nevertheless it is the study of the foraging behaviour of the latter that is currently thought by many to be most likely to further the understanding of predator–prey dynamics (e.g. Hassell, 1978). Thus it is relevant to ask: should predators continue to be regarded as parasitoids with complex life cycles? The simpler life cycles of parasitoids, for good pragmatic reasons, have made them more attractive as experimental animals than predators. However, this does not justify equating predation and parasitism. One objective of this book is to show that although these processes share some features they also differ sufficiently to warrant not lumping predators with parasitoids. Indeed it is only by studying predators that we are likely to determine if and how predator–prey dynamics differ from parasitoid–host dynamics.

The many studies on parasitoid–host interactions have generated a large body of theory built mainly around the concept that populations are regulated by density-dependent processes such as parasitoid–host and predator–prey interactions. In the context of biological control this body of theory has been used (1) to explain when and how natural enemies regulate their host or prey populations and (2) to develop techniques for detecting and evaluating the effectiveness of natural enemies. In general, however, attempts to use the theory of population dynamics to highlight the attribute(s) of effective biocontrol agents have been unsuccessful (Luck, 1990). However, most models of host–parasitoid interactions assume that hosts are equally acceptable regardless of their quality and the rate at which they are encountered. A parasitoid in parasitizing a host is making a decision that will decide its potential fitness. Therefore, more attention should be paid to the behaviour of the natural enemy, in particular its decision-making (Luck, 1990). Similarly, it has been suggested that the reason for the poor progress in developing a foraging theory for insect predators is that most studies have concentrated on the most voracious stage, the larva, rather than the adult. For a complete understanding, as with parasitoids, it is necessary to determine the behaviour that is most important in maximizing predator fitness, and this involves studying decision-making by adult predators (Ferran & Dixon, 1993). That is, another objective of this book is to collate the data in the literature in order to discover patterns and evolutionary constraints. This is used to reveal the factors that have shaped the foraging behaviour of ladybirds and to identify the attribute(s) of an effective biocontrol agent. Although currently not fashionable, such comparative studies of life histories have proved very effective in highlighting patterns and stimulating the study of processes in other groups of insects such as aphids (Dixon, 1998).

Ladybirds have supplied both the most famous case of biological control and many cases in which they have not proved effective. A comparison of the successful and unsuccessful attempts to use ladybirds in biological control, as has been suggested for parasitoids (Luck, 1990), is likely to complement the behavioural study of predator–prey interactions and give a better understanding of why some attempts to use ladybirds have been successful and others not. In addition, by highlighting the differences between predation and parasitism it might reveal the common attribute(s) of predators and parasitoids that determines the abundance of their prey and hosts, and so achieve a better and more general understanding of 'predator'–prey dynamics. This is the final objective of this book. In addition to paraphrase Medawar (1965) it is hoped that this analysis of ladybirds as predators will go some way to relieving us of the ever-increasing burden of the singular instances, the tyranny of the particular. The enormous ballast of factual information, so far from being just about to sink us, is used to reveal patterns and processes so that we need no longer record the fall of every apple.

2

Basic biology and structure

Ladybirds are beetles (Coleoptera), an ancient and very successful group of insects, which evolved in the Lower Permian some 280 million years ago. By the end of the Jurassic all the modern superfamilies of Coleoptera were established as distinct lineages. The phylogeny of the Coleoptera, based mainly on comparative studies of modern beetle morphology, has tended to place the ladybirds (Coccinellidae) among the more advanced beetles (Crowson, 1981). However, a molecular phylogeny based on mitochondrial cytochrome oxidase indicates that the ladybirds may be closer to the more primitive members of Coleoptera, the ground beetles (Carabidae), than previously thought (Howland & Hewitt, 1995). That is, there is some uncertainty about the precise phylogenetic position of the ladybirds within the Coleoptera. However, there appears no doubt that the ladybirds constitute a very distinct monophyletic family, the Coccinellidae, which includes 4500 species world wide.

The family name, Coccinellidae, means clothed in scarlet, and refers to the predominant scarlet colouration of the wing covers of ladybird beetles. Although an accurate description of most of the Coccinellinae it does not apply to the majority of the ladybirds, which are dull in colour. That is, the origin of the family name is to be found in the relatively large and spectacularly scarlet Coccinellinae, which were the first species to be collected and described. The currently accepted phylogenetic relation between the various subfamilies of the Coccinellidae is given in Fig. 2.1. Of the subfamilies both the Coccinellinae and Epilachninae contain species that feed on fungi or higher plants. Of the predatory species most feed on either aphids or coccids, with a few feeding on both types of prey. In addition some species feed on mites (Putman, 1955), adelgids (Delucchi, 1954; Pope, 1973), aleyrodids (Heinz &

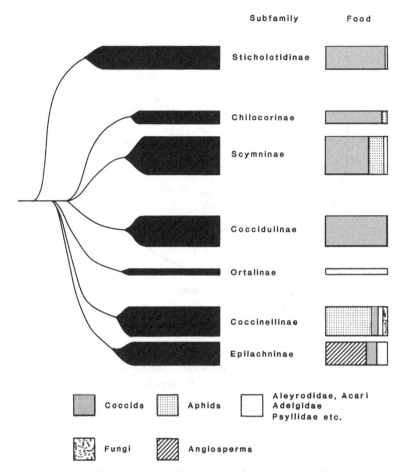

Fig. 2.1. Dendrogram showing the phylogenetic relationships of the subfamilies of Coccinellidae, with on the right an indication of the main foods
of each of the subfamilies. (After Hodek & Honěk, 1996.)

Zalom, 1996), ants (Harris, 1921; Pope & Lawrence, 1990), chrysomelid
larvae (Elliot & de Little, 1980), cicadellids (Ghorpade, 1979), pentatomids
(Subramanyam, 1925), phylloxera (Pope, 1973), mycophagous ladybirds
(Camargo, 1937) and psyllids (Booth, 1997; Chazeau *et al.*, 1991), and one
species, *Cleidostethus meliponae*, which is wingless and blind, has only been
collected from the nests of a meliponine bee, *Melipona alinderi*, in East
Africa (Salt, 1920). The food of ladybirds in a particular region is likely to
reflect the faunal composition of the potential prey in that area (cf.
Chapter 3). For example, in Central Europe aphids are more numerous in
terms of species than coccids, and a greater proportion of the ladybirds

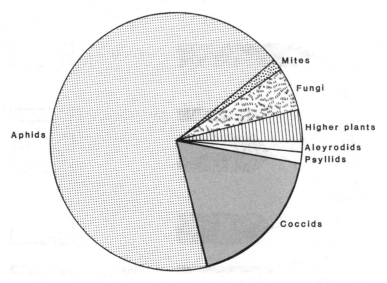

Fig. 2.2. Pie diagram indicating the main foods of ladybirds in Europe. (After Klausnitzer, 1993.)

there specialize on aphids than on coccids (Fig. 2.2). In other regions where there are few species of aphids, like Australia, most of the indigenous ladybirds are likely to feed on more species-diverse groups, such as coccids or psyllids (Hales, 1979).

LIFE CYCLE

In common with that of other holometabolous insects the ladybird life cycle starts with an egg, which hatches to give rise to a larva that goes through four instars, and then pupates and metomorphoses into an adult (Figs. 2.3, 2.4). The constant and same number of instars in aphidophagous and coccidophagous ladybirds is surprising. In the aphidophagous species fast development appears to be adaptive (p. 67) and this could be achieved by reducing the number of instars. However, of the many studies done on the development of ladybirds only one species is recorded completing its development in three rather than the usual four instars. Surprisingly it is not an aphidophagous but a coccidophagous species (McKenzie, 1932). In addition, as other groups of beetles, like the Silphidae, have three larval instars (Růžička, 1992), it does not appear that the number of instars is constrained by phylogeny.

A striking difference between aphidophagous and coccidophagous ladybirds is that the former generally lay eggs in clusters whereas the

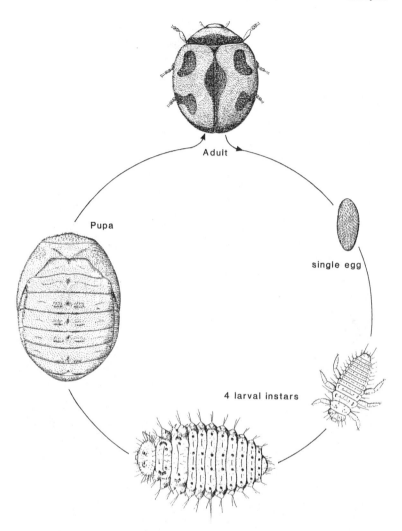

Fig. 2.3. Life cycle of a coccidophagous species of ladybird, *Rodolia cardinalis*.

latter lay their eggs singly. There are exceptions in that the aphidopha-gous *Platynaspis* lays its eggs singly, usually in crevices or rolled-up leaves where they are afforded some protection from the ants that attend this ladybird's prey (Völkl, 1995), and the coccidophagous *Orcus chalybeus* Boisd. lays eggs in clusters (Thompson, 1951). Some insight into why *O. chalybeus* is apparently unique in laying its eggs in clusters might indi-cate the adaptive significance of this behaviour in aphidophagous species.

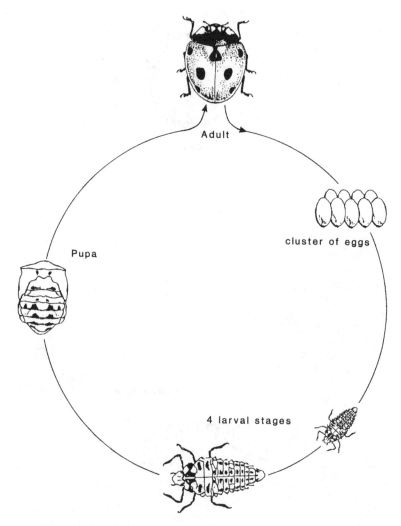

Fig. 2.4. Life cycle of an aphidophagous species of ladybird, *Coccinella septem-punctata*.

MORPHOLOGY

The basic structure of an adult ladybird is illustrated in Fig. 2.5. Amongst the predatory ladybirds there appears to be very little variation in the shapes of the adults. A logarithmic plot of breadth against length for 55 species (Fig. 2.6) indicates a directly proportional relationship, in which a doubling of length is very closely associated with a doubling in breadth and accounts for 93% of the variation observed between species.

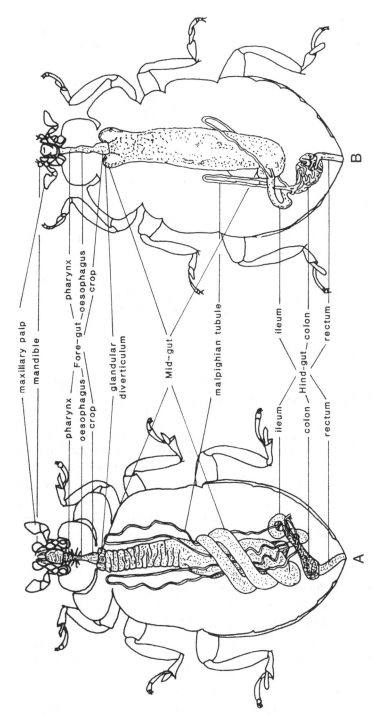

Fig. 2.5. Alimentary canal and basic structure of an adult herbivorous (A) and carnivorous (B) ladybird. (After Pradhan 1936, 1939.)

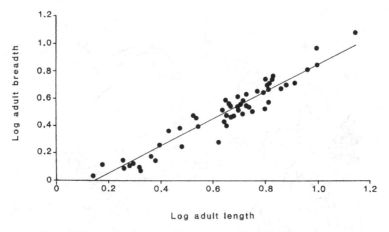

Fig. 2.6. The relationship between the logarithm of the breadth and the logarithm of the length of adults of 55 species of ladybird.

As with the life cycle, however, certain features of the morphology differ between beetles and these differences can be related to their way of life.

Mouthparts

In predatory ladybirds the mandibles tend to be powerfully built and are used to crush and tear their prey. The tooth at the tip of each mandible is either bi- or unidentate. The latter condition is characteristic of those coccidophagous ladybirds which use their mandibles to cut open the chitinous covering of their prey with a can-opener-like movement or prise their prey off the surface (Fig. 2.7(E); Samways & Wilson, 1988). However, in some species, like *Platynaspis*, *Stethorus* etc., each mandible comes to a sharp point with a groove running down its inner margin and functions like a hypodermic needle (Fig. 2.7(D)). These species usually grip an appendage of an aphid with their mandibles, inject enzymes and suck out the digested body tissues. The herbivorous ladybirds use their mandibles to scrape tissue from the surface of leaves, which is reflected in the shape of each mandible, the inner cutting edge of which extends from the apical tooth to the molar region (Fig 2.7(A)). Those that feed on fungal spores, like *Tytthaspis* spp., have comb-like structures on their mandibles with which they gather fungal spores (Fig. 2.7(B), (C)) (Minelli & Pasqual, 1977; Ricci, 1979, 1982; Samways *et al.*, 1997; Samways & Wilson, 1988).

Closely associated with the mandibles are the maxillae and maxillary palps. The latter are thought to play a role in prey recognition (Kesten, 1969; Nakamuta, 1985*a*). Amputation of the maxillary palps

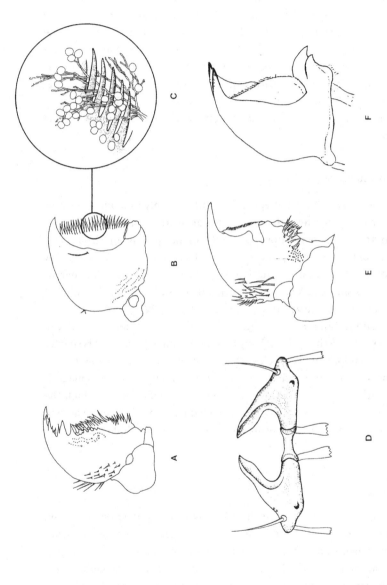

Fig. 2.7. Ladybird mandibles: (A) phytophagous species, *Epilachna reticulata*; (B) mycophagous species, *Tythaspis sedecimpunctata*; (C) magnified portion of the inner edge of a mandible of (B) showing the comb-like structure this species uses to harvest fungal spores; (D) hypodermic-like mandibles of *Platynaspis*; (E) coccidophagous species, *Chilocorus nigritus*; (F) aphidophagous species, *Coccinella septempunctata*. (After Pradhan, 1938; Ricci, 1979, 1982; (Samways *et al.*, 1997.)

results in a decrease in the efficiency of prey capture by approximately 40% in *Coccinella septempunctata brucki* (Nakamuta, 1985b). The palps appear to differ greatly in size and sensory capability in different groups of ladybirds. Although the sample size is small, and the different sizes of the beetles could be a confounding factor, there appears to be some support for the suggestion that the size of the maxillary palps and the number of sensory receptors they bear are largest in aphidophagous and smallest in phytophagous species. It has been suggested that this trend is associated with the speed with which a ladybird has to respond to the presence of prey, with aphids less easily caught than coccids, and plants even more easily located because of their size and immobility (Barbier *et al.*, 1996).

Alimentary canal

The alimentary canal appears to differ mainly in length (Fig. 2.5). Herbivorous species tend to have a gut that relative to their size is twice the length of that of predatory species (Pradhan, 1936, 1939). This reflects the need of herbivorous species to process large quantities of low-quality food, of which they only assimilate 23% of the energy content, whereas carnivorous species process smaller quantities of higher quality food of which they assimilate 77% (Brafield & Llewellyn, 1982).

Again although few species have been studied there is some evidence to indicate that the gut length in coccidophagous species is shorter than in aphidophagous species. This has been attributed to the higher nutritional value of coccids and the lower voracity of coccidophagous species (Iperti *et al.*, 1977). However, a more detailed study, in which the confounding effect of body size is corrected for, is needed before accepting that gut length is shorter in coccid-feeding than aphid-feeding ladybirds.

Legs

The larvae like the adults have legs, which in most species are used for locomotion. The legs of the larvae of aphidophagous species are notably longer than those of coccidophagous species (p. 68). This is possibly associated with the former feeding on generally more mobile prey than the latter. In some species, like the two-spot ladybird *Adalia bipunctata* the larvae have a large glandular swelling around their anus, the pygopodium, which they periodically apply to the substrate. This foot in their rump possibly serves to give them added adhesion on smooth surfaces and is of particular advantage to tree-dwelling species in which dislodgement is particularly hazardous. In the larvae of the whitefly predator, *Clitostethus arcuatus*, locomotion is similar to that observed in the inch worm. The legs do not appear to have an ambulatorial function. The pygopodium is the main point of support, the legs claw the substrate and stretch the body like a spring and then the mouthparts are used to establish a new point of attachment before the pygopodium's hold on the substrate is released and the body arches forward and the process is repeated (Ricci & Cappelletti, 1988).

DEVELOPMENT

Although at a particular temperature ladybirds specializing on aphids take considerably less time to develop from egg to adult than those feeding on coccids (Chapter 4), nevertheless the proportion of the total time they spend in the egg, larval and pupal stages is very similar (Fig. 2.8). Even the limited data available for three genera of phytophagous ladybirds (*Subcoccinella, Henosepilachna* and *Epilachna*) and one mycophagous ladybird, *Leptothea galbula*, indicate that the proportion of time spent in the egg, larval and pupal stages is 0.18, 0.62 and 0.20, and 0.13, 0.64 and 0.23 respectively, which is very similar to that recorded for the predatory ladybirds (cf. Fig. 2.8; Ali, 1979; Anderson, 1980; Nakamuta, 1987; Fan *et al.*, 1992). Developmental time is very dependent on both temperature and food quality. Once this is widely appreciated, and the stage of development recorded more frequently at higher temperatures, then it is likely that each ladybird will be shown to spend approximately the same proportion of its total developmental time in each of the developmental stages. That is, the ratios of the time spent in the different stages do not alter with temperature; this is referred to as rate isomorphy. Interestingly rate isomorphy is independent of species and rearing temperature within a species (Jarosik *et al.*, 2000). Although there have been many claims that insects reared under more natural fluctuating temperature conditions develop faster than those reared under constant temperature conditions, this has rarely been rigorously tested. In the case of the ladybird *Scymnus hoffmani* there is no significant difference in the length of larval development when reared under fluctuating and constant temperatures of equivalent value in day degrees (D°) (Ding-Xin & Zhong-Wen, 1982). Similarly, an analysis of the results of Gawande (1966) for *Menochilus sexmaculatus* revealed no difference.

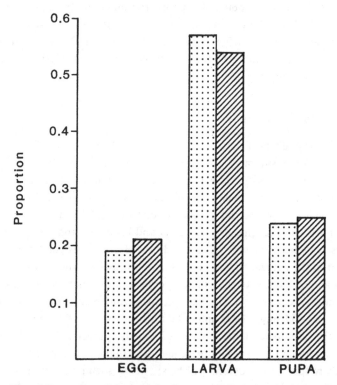

Fig. 2.8. Proportion of the total developmental time spent in the egg, larval and pupal stages in aphidophagous and coccidophagous species of lady-birds. (Data from: Aguilera, 1995; Aguilera & Diaz, 1983; Ahmad, 1970; Bain *et al.*, 1994; Benham & Muggleton, 1970; Brettell, 1964; Brown, 1972; Butler, 1982; Butler & Dickerson, 1972; Campbell *et al.*, 1980; Carrejo *et al.*, 1991; Chakrabarti *et al.*, 1988, 1995; Chazeau *et al.*, 1991; de Fluiter, 1939; Geyer, 1947*a*; Greathead & Pope, 1977; Gurney & Hussey, 1970; Hecht, 1936; Jalali & Singh, 1989*b*; Kawauchi, 1983; Meyerdirk, 1983; Michels & Bateman, 1986; Michels & Behle, 1991; Miller & LaMana, 1995; Moursi & Kamal, 1946; Nadel & Biron, 1964; Naranjo *et al.*, 1990; Okrouhlá *et al.*, 1983; Pantyukhov, 1965, 1968; Schanderl *et al.*, 1985; Sharma *et al.*, 1990; Simpson & Burkhardt, 1960; Varma *et al.*, 1993; Zhao & Wong, 1987.)

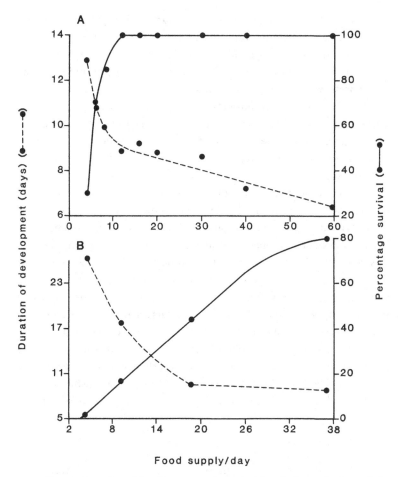

Fig. 2.9. Duration of development and survival in relation to food supply in (A) *Propylea japonica* and (B) *Adalia bipunctata*. (After Dimetry, 1976; Kawauchi, 1979.)

SURVIVAL

In laboratory studies the more food supplied to larvae per day the more of them survive (S) to maturity and the shorter their development (D). There is a strong correlation between percentage survival and duration of larval development in *Adalia decempunctata* (Dixon, 1959):

$$S = 154 - 6.3 D \qquad (2.1)$$

This relationship accounts for 96% of the variation. Other species of ladybird show similar very close associations between survival and duration of development (Fig. 2.9).

In the field very high levels of mortality during development have been recorded (p. 148), and it is often attributed to starvation. In the case of insect herbivores in the field it is assumed that slow growth may translate into an increase in enemy-inflicted mortality either by increasing the herbivores' 'window of vulnerability', as they are exposed to natural enemies for longer, or because they also spend longer feeding each day. This is referred to as the slow-growth-high-mortality hypothesis (Clancy & Price, 1987). However, tests of this hypothesis indicate that prolonged development does not always result in increased parasitism (e.g. Benrey & Denno, 1997).

That slow-growth-high-mortality occurs in herbivores like aphids (Dixon, 1998) and carnivores like ladybirds indicates this phenomenon is not restricted to a particular trophic level. That it occurs in the laboratory in the absence of arthropod natural enemies tends to question their role in slow-growth-high-mortality. Although pathogens may be responsible the condition of the ladybird larvae that die makes this unlikely. One possible explanation is that although the rate of growth of larvae is adjusted to food supply (p. 39), nevertheless individuals are inherently fast or slow developers. When food is short only the slow developers can acquire sufficient food to support their growth. With increasing food supply a greater proportion of individuals are able to sustain their inherent growth rate, and so survive. Overall this would give the linear relationship between survival and food supply observed. This needs to be tested experimentally.

REPRODUCTION

Reproductive output in most insects, including ladybirds, starts and reaches a maximum early in adult life, and then declines (Fig. 2.10). That is, they show a triangular fecundity function the shape of which is dependent on food supply/quality and temperature, particularly the latter. When temperature is high the peak in egg production is generally higher and occurs earlier, and the subsequent decline in output is faster than when temperature is low (Ding-Xin & Zhong-Wen, 1987). This results in the fecundity function tending towards a right-angled triangular shape at high temperatures and an equilateral triangular shape at low temperatures.

Fecundity and longevity

Why insects show a triangular fecundity function is still being debated. It is usually assumed that insects maintain a constant reproduc-

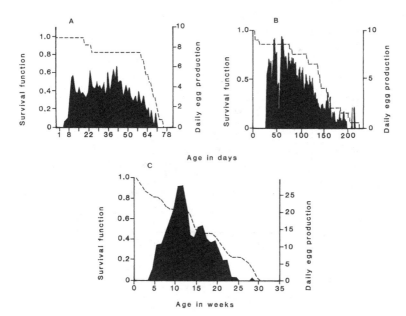

Fig. 2.10. Survival (- - - -) and egg production (■) of (A) *Scymnus hoffmani* (B) *Coelophora mulsanti* and (C) *Scymnus frontalis*. (After Gibson *et al.*, 1992; Sallée & Chazeau, 1985.)

tive output per unit time throughout adult life and that it is mortality that shapes the fecundity function (Stearns & Koella, 1986). Certainly in the field adult mortality is going to be a major factor in determining realized fecundity. Even if the probability of dying does not change with age, the probability of being alive to reproduce is much greater early than late in adult life. Thus there is a clear advantage in early reproduction. Any change that results in even more of the reproductive output being generated in early adult life, even if this lowers the probability of long-term survival, would also be advantageous. That is, there is likely to be a fecundity–longevity trade-off, of which ageing is a manifestation.

What evidence is there for such a trade-off? Within a species of ladybird, unmated individuals lay fewer eggs and live longer than mated individuals (p. 75), and the short-lived aphidophagous ladybirds have a higher fecundity than the longer-lived coccidophagous ladybirds (p. 78). Similar-sized adults of *Adalia bipunctata* fed a high-quality diet lay more eggs and survive for a shorter period of time than those fed an excess of poor quality food (Kariluoto, 1980). That is, they appear to show a trade-off between fecundity and longevity. In addition, in the congenial conditions of a laboratory the triangular fecundity function is determined by beetles

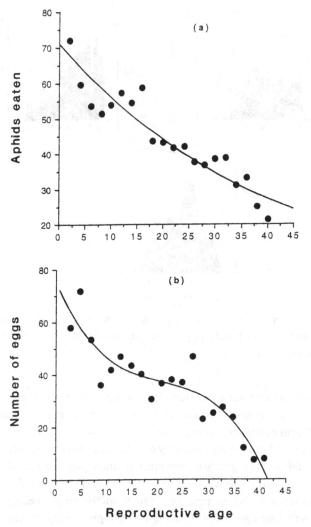

Fig. 2.11. The number of (a) aphids eaten, (b) eggs laid and (c) size of egg clusters in relation to age (in days) in *Menochilus sexmaculatus*. (Agarwala, personal communication.)

producing more eggs per unit time in early than in late adult life. In *Menochilus sexmaculatus* the decline in reproductive output with age is closely associated with a reduction in the number of eggs laid in each cluster and a reduction in the daily consumption of aphids (Fig. 2.11). Of particular relevance to the idea that ageing is important in shaping the fecundity function is that Figs. 2.11(a) and (b) indicate that old adults appear to be nearly three times less efficient at converting aphid biomass into eggs than young adults. It would be interesting to know how much of

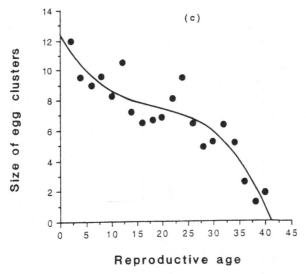

Fig. 2.11. (cont.)

this is due to the early utilization of stored resources (fat) to boost egg production and how much due to the concomitant affects of ageing on the efficiency of assimilation and conversion. Although the underlying mechanism still remains to be determined there is nevertheless good evidence of a trade-off between fecundity and longevity in ladybirds and this could have been important in shaping their triangular fecundity function.

Effect of food supply on egg and cluster size

Although there has been much interest in insect reproductive tactics, in particular how many eggs should be laid in a patch (e.g. Parker & Courtney, 1984; Godfray, 1987; Mangel, 1987), little attention has been given to the factors that determine egg and cluster size in predatory insects. Aphidophagous ladybirds lay their eggs in clusters, which vary in size both within and between species. Coccidophagous species tend to lay eggs singly and virtually nothing else is known about their egg-laying tactics. It has been suggested that egg size in aphidophagous ladybirds is constrained by the minimum size at which first instar larvae can capture active prey (Stewart et al., 1991b). This leads to the prediction that within a species of ladybird, as in birds (cf. Lack, 1986), egg size is likely to be the least variable reproductive trait and cluster size the most variable. This hypothesis was tested by Dixon & Guo (1993) by determining the direct and indirect affects of aphid abundance on egg and cluster size in the seven-spot ladybird, Coccinella septempunctata.

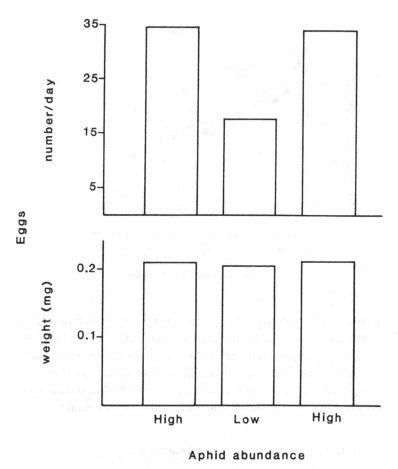

Fig. 2.12. The average weight of an egg and the number laid per day by *Coccinella septempunctata* when supplied first with a high, then a low and finally a high number of aphids per day. (After Dixon & Guo, 1993.)

Direct effects

Both the number of eggs laid per day and the size of individual clusters of eggs are greatly affected by the availability of food. When food is supplied in excess beetles lay twice as many eggs per day and in larger clusters than when the same beetles are fed at a third of this rate. In marked contrast the average size of the eggs laid by individual females is not affected by varying their food supply (Fig. 2.12). Thus, as predicted, when food supply varies cluster size and number of eggs produced per day are more likely to vary than egg size.

Indirect effects

When fed an excess of aphids individual beetles consistently lay either small or large clusters of eggs, with large individuals laying significantly larger clusters than small individuals (Fig. 2.13(A)). Large beetles also have more ovarioles than small beetles (Fig. 2.13(C); Rhamhalinghan, 1985) and beetles with a large number of ovarioles lay larger clusters of eggs than beetles with few ovarioles (Fig. 2.13(B)). Therefore, not surprisingly, there is a relationship between the residuals about the relationship between cluster size and adult size and the ovariole number of the individual beetles (Fig. 2.13(D)). That is, variation in ovariole number in adults of the same size accounts for some of the variation observed in the relationship between egg cluster size and adult size (Fig. 2.13(A)).

Several studies have shown that egg production in ladybirds is determined by the availability of prey (p. 158). However, the availability of prey during the development of ladybirds, through its effect on adult size (p. 37), can also affect egg production. Even when prey is abundant small individuals are less fecund than large individuals. This is likely to be determined by the small number of ovarioles in the gonads of small beetles and the likelihood that the speed with which eggs are produced by each ovariole is rate-limited. Thus potential egg production is affected by both aphid abundance experienced by a beetle during its larval development and that available to it as an adult (Fig. 2.14). In addition, as prey also varies in quality the nature of the prey can also determine egg production. This is well demonstrated by studies on *A. bipunctata* and *M. sexmaculatus* by Blackman (1965, 1967), Rajamohan & Jayaraj (1973) and Kariluoto (1980).

Interspecific relationships

Large species of both aphidophagous and coccidophagous ladybirds, like other animals, appear to invest less relative to their body weight in individual offspring than small species (p. 61). In addition, aphidophagous ladybirds lay their eggs in clusters the size of which is dependent on the number of ovarioles (*Ov*) in the gonads of each species (Stewart *et al.*, 1991a). Assuming that the proportion of the body made up of gonads is the same for all aphidophagous ladybirds then the effect of the interspecific variation in ovariole number on egg weight (*E*) can be corrected for by dividing adult weight (*Wa*) by ovariole number (*Wa/Ov*). The relationship between the logarithm of egg weight (*E*) and the logarithm of *Wa/Ov* for eight species is described by:

$$\log E = 0.83 \log Wa/Ov - 0.44 \qquad (2.2)$$

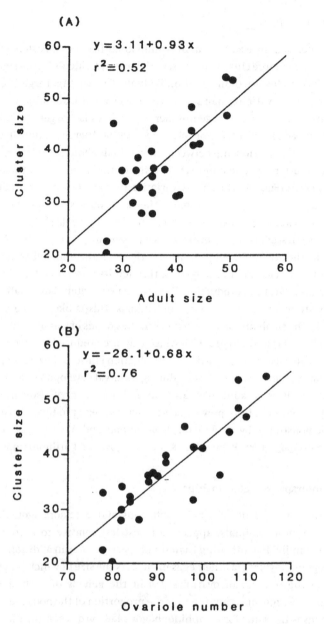

Fig. 2.13. Egg cluster size in relation to (A) adult size and (B) ovariole number, ovariole number in relation to adult size (C) and the residuals in relationship (A) in relation to ovariole number (D) in *Coccinella septempunctata*. (After Dixon & Guo, 1993.)

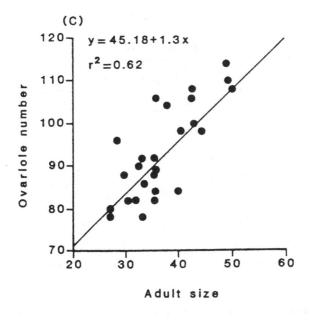

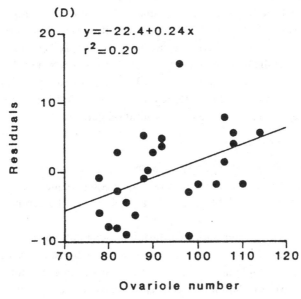

Fig. 2.13. (*cont.*)

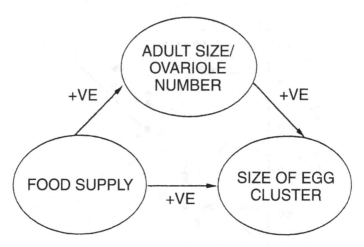

Fig. 2.14. Schematic diagram of the direct and indirect effects of food supply on the size of egg clusters laid by ladybird beetles. (After Dixon & Guo, 1993.)

This indicates that a species with few ovarioles lays larger eggs than a similar sized species with many ovarioles. The allometric coefficient is not significantly less than 1.0. Thus egg weight increases in direct proportion to body weight when the ovariole number is kept constant. Therefore, it is likely the gonads make up a fixed proportion of the body weight. If the reproductive biomass is defined as the ovariole number multiplied by egg weight then there is a very strong positive correlation between reproductive biomass (R) and adult weight:

$$\log R = 1.05 \log Wa - 0.44 \tag{2.3}$$

In this relationship the allometric coefficient (1.05) is not significantly different from 1.0. This indicates, not unexpectedly (cf. equation 2.2), that doubling of adult weight results in a doubling of the reproductive biomass.

There is little information on the reproductive rate (Rr), i.e. biomass of eggs produced per day by different species of ladybirds. The values for four species of a wide range of adult weights (9.8 to 47.8 mg) indicate a strong positive correlation between reproductive rate and adult weight (Stewart et al., 1991b):

$$\log Rr = -0.7 + 0.97 \log Wa \tag{2.4}$$

That is, doubling adult weight possibly also results in a doubling of the potential reproductive rate in aphidophagous ladybirds.

Developmental time (D) in days at 20 °C shows a good positive correlation with the ratio of adult to egg weights (Wa/E) such that relatively small eggs take longer to develop to the adult stage. The relationship is:

$$D = 0.07\,(Wa/E) + 11.24 \qquad\qquad (2.5)$$

Thus, as predicted, the trade-off between egg number and weight has a marked affect on developmental time. Species with proportionally small eggs take longer to complete their development than those with proportionally large eggs.

As indicated above there is little data on adult weight, ovariole number and developmental time for particular species. Therefore, there is an urgent need to collect more data with the specific objective to test the generality of the above relationships. The fact that aphids show similar relationships tends to support the notion of such trade-offs (Dixon, 1998). The very little information available on coccidophagous ladybirds (Magro, 1997) also tends to support the notion that the reproductive biomass of these species also makes up a fixed proportion of the adult body mass. However, the proportion in coccidophagous species is half of what it is in aphidophagous species. Similarly the reproductive rate (mg of eggs produced per day) of coccidophagous species appears to be only a third of that of similar-sized aphidophagous ladybirds. This lends further support to the message of Chapter 3 that aphidophagous ladybirds invest proportionally more in reproduction than coccidophagous species and as a consequence may have a shorter adult life than the latter (p. 76).

OVERWINTERING

The life cycle of a ladybird is seen mainly in terms of the development from egg to adult. However, during late autumn and winter in temperate regions food is scarce and weather unfavourable for ladybirds and their prey. The short day conditions in autumn induce ladybirds developing then, and adults surviving from earlier generations, to lay down fat reserves (Hodek & Cerkasov, 1960 ; Hodek *et al.*, 1984*b*) and seek overwintering sites. In addition, beetles may respond to other cues heralding the onset of autumn. For example, *Semiadalia undecimnotata* that feed on aphids reared on senescent plants regress their ovaries and accumulate fat reserves, whereas those fed aphids from young plants remain reproductively active (Rolley *et al.*, 1974); and *Hippodamia convergens* fed summer parthenogenetic aphids are reproductively more active than those fed the autumnal sexual forms of the same aphid (Wipperfürth *et al.*, 1987). In a few cases hibernation involves migration to specific sites. For example, there are enormous numbers of *H. convergens* that overwinter in the Sierra Nevada mountains and of *S. undecimnotata* in the Alpes de Haute Provence (Hagen, 1966; Iperti & Buscarlet, 1986).

Although many studies have been carried out on various aspects of overwintering the ecology of ladybirds during this period is poorly understood. Survival appears to be better in some sites than in others (Honěk, 1989) and beetles that mature and accumulate fat reserves early are the most likely to locate these sites. Thus there would appear to be a clear advantage to an early switch from a reproductive to an overwintering mode; however, this curtails further reproduction. The marked variation in the time of onset of autumn and abundance of prey in late summer from year to year will differentially affect the fitness of beetles that switch early or late. This uncertainty is likely to sustain a genetic-based polymorphism in the time of the switch.

It is likely that survival over winter is dependent on the amount of metabolic reserves a ladybird can accumulate prior to entering diapause (Hodek & Cerkasov, 1963; Tauber *et al.*, 1986; Danks, 1987). However, it is difficult to obtain a direct measure of the fat reserves of living beetles. In addition, size is also likely to prove a confounding factor. Individuals of *C. septempunctata* that are still foraging late in autumn are significantly lighter than those that are already hibernating. If the foraging ladybirds are prevented from foraging then only 28% of them survive over winter compared to 91% of those that have entered diapause. Although a higher percentage of the larger than the smaller individuals in each category survived, the effect of size was considerably less than the effect of category (Barron & Wilson, 1998). Another study, in which specimens of *C. septempunctata* were collected from hibernation sites, classified according to size and overwintered in artificial hibernacula outdoors, revealed that survival is not associated with gender or consistently with body size (Honěk, 1997). Interestingly, in this context the females contain more fat per unit weight than males (Zhou *et al.*, 1995). In the Czech Republic individuals of *C. septempunctata* that hibernate at high altitudes have significantly larger fat reserves than those that hibernate at lower altitudes. Interestingly, adults hibernating at low altitudes in the Czech Republic and the United Kingdom use up 50% of their fat reserves, whereas those hibernating at high altitudes in the Czech Republic use only 30%. Although it is uncertain how much of this fat is available to beetles for metabolism these results indicate that adults hibernating at high altitudes, where temperatures are low, may be able conserve more of their fat reserves (Zhou *et al.*, 1995). The greater the fat reserves on emerging from hibernation the greater the chance of surviving to locate prey (Mills,

1981*a*). In addition they may also use these reserves to enhance their egg production in spring.

What evidence is there that overwintering ladybirds use their fat reserves to enhance their reproductive status in spring, i.e. do they show generation-specific strategies? *H. convergens* emerging from overwintering have a shorter pre-oviposition period and consume fewer aphids during this period than subsequent generations. Over the first three generations the length of the pre-oviposition period and the number of aphids consumed during this period increases with each generation (Hagen & Sluss, 1966). Similarly, *M. sexmaculatus* collected from hibernation sites and kept in an insectary for four generations show a decrease in average fecundity and average period of oviposition in each subsequent generation from 1948 to 154 eggs and 91 to 34 days respectively (Hukusima & Kouyama, 1974). In *Propylea japonica* fecundity is also highest in the generation that emerges from hibernation (Hukusima & Komada, 1972). Although there is no difference in the fecundity of *A. bipunctata* reared under long day conditions in a laboratory and those collected from hibernation sites and kept under the same conditions, individuals from hibernation sites initially had a higher rate of egg production than those reared in the laboratory (Hemptinne, personal communication). El-Hariri (1966) also reports that this species starts developing one egg in each ovariole before emerging from hibernation, unlike *C. septempunctata* and *Propylea quatuordecimpunctata* which have to feed before maturing eggs.

These insectary and laboratory observations suggest the existence of generation specific strategies as have been reported in other insects, in particular aphids (Dixon, 1998). The fact that a similar phenomenon has been recorded in several species of ladybird further supports this contention. However, these trends in life history traits could simply be a consequence of changes in rearing conditions, i.e. environmentally induced phenotypic plasticity. Other authors suggest it might be due to inbreeding depression (Hodek & Cerkasov, 1960). Therefore, there is now a need to test specifically whether ladybirds have generation-specific strategies. If they do then it will also be important to determine the extent to which these strategies are programmed and/or cued by environmental stimuli, and their adaptive significance.

DEFENCE

When molested or disturbed adults and larvae of ladybirds often feign death and/or exude a yellowish or reddish fluid from tibio-femoral articulations in the case of adults and dorsal glands in the case of larvae, which is called reflex bleeding. As the exuded fluid is bitter, toxic and often has a strong smell it has long been regarded as a defence against insect or vertebrate predators. As is the case with many other aposematic insects the smell is due, at least in part, to 2-isopropyl-3-methoxy-pyrazine (Al Abassi *et al.*, 1998). That is, these insects are likely to gain additional protection by looking and smelling similarly. This is referred to as Müllerian mimicry (Rothschild, 1961; Moore & Brown, 1981; Moore *et al.*, 1990). Interestingly, the dull-coloured ladybirds are virtually devoid of pyrazines, but do reflex bleed, in the case of *Rodolia cardinalis* prolifically so, which results in collecting equipment becoming stained a cardinal colour (Hemptinne, personal communication).

The chemical defence system of ladybirds is mainly based on repellent and in some cases toxic alkaloids, some of which are known to be autogenously produced (Tursch *et al.*, 1976; Ayr & Browne, 1977; Jones & Blum, 1983). Alkaloids have been reported in species in only four subfamilies of ladybirds: Coccinellinae, Chilocorinae, Epilachninae and Scymninae, and most of the studies have been on species of the strikingly aposematic and largest subfamily, the Coccinellinae. It is highly likely that species in other subfamilies, so far not studied, will also be shown to contain alkaloids (Fig. 2.15). Interestingly, these alkaloids consist of a few homologous building-blocks derived from the amination of simple fatty acids. The amazing diversity of alkaloids in ladybirds indicates they are among the pioneers of combinatorial chemistry (Schröder *et al.*, 1998). Although each species often contains a mixture of alkaloids there is usually one major alkaloid, e.g. adaline in ladybirds of the genus *Adalia* and coccinelline in those of the genus *Coccinella*. In *Exochomus quadripustulatus* the concentration of exochomine appears to be the same in all the life stages, from egg through to adult. In addition the concentration in the haemolymph and reflex fluid is very similar, which indicates that they may truly reflex bleed with the fluid exuded being haemolymph minus blood cells. A better understanding of the distribution and function of the alkaloids both within and between genera and subfamilies will give a chemotaxonomic view into the phylogeny of the Coccinellidae (Daloze *et al.*, 1995). For example, the major alkaloid in *Coccinella magnifica* is hippodamine not coccinelline (Lognay, personal communication). Although *C. magnifica* is outwardly very similar to *C. septempunctata* its having a very different alkaloid challenges it being assigned to the same genus.

There is great variation between individuals of the two-spot ladybird *Adalia bipunctata* in the amount of reflex blood produced. If defence is energetically expensive then one would expect a negative association

SUBFAMILY   ALKALOIDS

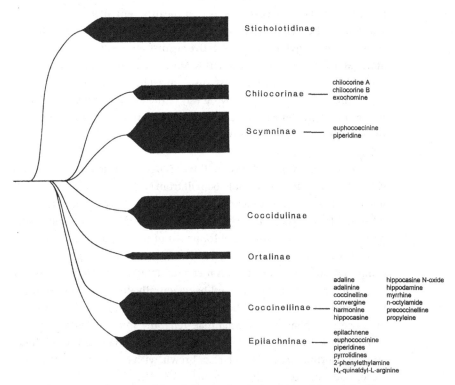

Fig. 2.15. Dendrogram of the phylogenetic relationships of the subfamilies of Coccinellidae, with on the right the alkaloids that have been recorded from species in each of the subfamilies.

between the quantity of adaline in the reflex blood, body size and growth rate. In the two-spot ladybird there appears to be no clear trade-off (Holloway *et al.*, 1993). Similarly, the larvae of *Harmonia axyridis*, which are reflex bled at regular intervals during their development only *tend* to take longer to develop and grow to a smaller adult size than those that are not reflex bled (Grill & Moore, 1998). The higher concentration of alkaloids in the reflex fluid of females than of males reported for the two-spot ladybird (de Jong *et al.*, 1991) may indicate that a higher concentration is needed to supply the eggs with their alkaloid defence or that ovipositing females are at a greater risk of attack by predators.

Two-spot ladybirds have been regarded as Batesian mimics of the chemically well-defended seven-spot ladybird *C. septempunctata* (Brakefield, 1985; Marples, 1990). As the two-spot ladybirds produce large quantities of reflex fluid rich in adaline they clearly do not only depend on their aposematic resemblance to the seven-spot ladybird for protection (de Jong *et al.*, 1991). It is likely that the different alkaloids have been selected for their effectiveness against specific predators. Coccinelline in the seven-spot ladybird is very effective against some but not all bird predators (Marples *et al.*, 1989; Majerus & Majerus, 1998) whereas adaline is probably more effective against insect predators (Agarwala & Dixon, 1992). The large size and sun-loving way of life of the seven-spot ladybird may put it at greater risk from birds than from insect predators, whereas the reverse is probably true for the smaller and more secretive two-spot ladybird.

As it is likely that all coccinellids are chemically defended then provided they look alike they would benefit from Müllerian mimicry. As each potential predator has to learn that a particularly coloured ladybird is unpalatable the risk to a ladybird is reduced if it shares the same colour pattern with another unpalatable species of prey. That is, as more and more species are drawn into a Müllerian complex, the power of the complex to draw in other species increases. In addition, ladybirds are all very similar morphologically and behaviourally. In such circumstances the theory predicts that all ladybirds should look similar. However, the mainly aphidophagous Coccinellinae generally sport colour patterns of two strongly contrasting colours and the coccidophagous Coccidulinae and Chilocorinae are mainly black or brown, with no bright patterning at all. In addition some aphidophagous ladybirds, like *Anisosticta novemdecimpunctata* and *Harmonia quadripunctata*, are cryptically coloured while overwintering but on emerging from hibernation, and in particular with the onset of feeding and *oviposition* develop the contrasting colouration more characteristic of these species. This indicates that the selection pressure changes during the course of a season (Majerus & Majerus, 1998).

The interpretation of this diversity of colour patterns in ladybirds has been mainly sought in terms of Batesian and Müllerian mimicry. After reviewing the extensive literature on this subject Majerus & Majerus (1998) concluded that not all the information necessary to explain the complex colour polymorphism shown by ladybirds is to hand and suggest it is time to give up looking for a single unifying factor. In addition, the fact that the chemical defence of ladybirds is also effective against insect predators (Pasteels *et al.*, 1973; Agarwala & Dixon, 1992, 1993) warrants more consideration, especially as species like the seven-spot ladybird, which is known to be toxic to great tit chicks (*Parus major*) (Marples *et al.*, 1989), in ladybird plague years comprises a large proportion of the food fed to house martin (*Dolichon urbica*) chicks (Majerus & Majerus, 1998).

That is, the empirical evidence indicates that predation by birds is not the only factor that has shaped the evolution of what is popularly thought to be the characteristic colour patterns of ladybirds. It is possible that these colour patterns, which are likely to affect the efficiency with which they capture prey, may have an important role in the foraging behaviour of ladybirds (p. 104).

The ladybird beetle, *Pseudoscymnus kurohime*, which specializes on aphids that protect themselves from attack by predators and parasites by means of soldier aphids, protects its eggs by covering them with a faeces-like secretion (Arakaki, 1988). The very large ladybird *Megalocaria dillata* is also reported to protect its eggs by laying them on a spine or needle-like plant structure and then surrounding them with a colourless sticky substance, which it secretes from the tip of its abdomen. This is reported to be an effective barrier against ants, spiders and ladybird larvae (Liu, 1933). Similarly the waxy covering of larvae of *Scymnus* affords them protection from ants and enables them to exploit ant-attended aphids (Völkl & Vohland, 1996). However, physical protection of the immature stages of ladybirds, especially the eggs, is rare. The eggs of aphidophagous ladybirds would appear to be particularly vulnerable to cannibalism or predation. Like the other stages in development, the eggs contain the same species-specific toxic alkaloids. Such a defence is of little value if the eggs are killed before their toxicity becomes apparent. Current studies indicate that the nature of the alkanes on the surface of an egg signal to a potential cannibal or predator the relative risk of attacking the egg (p. 183; Agarwala & Dixon, 1992; Hemptinne *et al.*, 2000b).

In addition to autogenously produced alkaloids ladybirds have also been reported sequestering the defensive chemicals of their prey. The seven-spot ladybird sequesters pyrrolizidine alkaloids (PA) when it feeds on the aphid *Aphis jacobaeae*, which is a specialist aphid adapted to feeding on PA-containing plants belonging to the genus *Senecio*, from which it sequesters the PA for its own defence (Fig. 2.16; Witte *et al.*, 1990). A ladybird, *Hyperaspis trifurcata*, that feeds on cochineal insect, *Dactylopius* spp., sequesters its prey's defensive chemical, the anthraquinone carminic acid, which is very active in deterring ants (Eisner *et al.*, 1994)

In conclusion, although all ladybirds share the same basic physiology and structure their different ways of life have resulted in differences in the structure of their mouthparts, and lengths of their alimentary canal and legs. Similarly, although there are big differences in the time it takes aphidophagous and coccidophagous ladybirds to complete their development the proportion of the total time spent in the egg, larval and pupal stages appears to be similar for

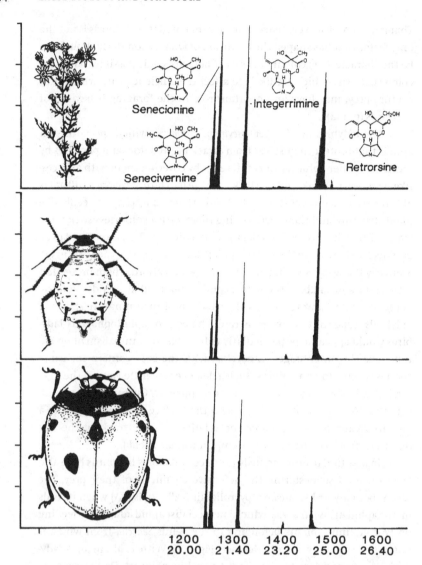

Fig. 2.16. The pyrrolizidine alkaloids present in *Senecio*, which are sequestered by its specialist aphid, *Aphis jacobaeae*, and seven-spot ladybirds, *Coccinella septempunctata*, which feed on this aphid. (After Witte *et al.*, 1990.)

these and other groups of ladybirds. The relationships between food supply and survival and reproduction, although based on data for a few mainly aphidophagous species, nevertheless also indicate that all groups of predaceous ladybirds show similar trends. However, the indication is that coccidophagous ladybirds invest less in gonads than aphidophagous species. The mass overwintering of several species of ladybirds is a very striking feature, but little is understood about this aspect of their ecology, in particular the role of fat reserves in survival and in reproduction on emergence from hibernation. Alkaloids are likely to be the common defence against predators of all ladybirds. However, the fact that not all ladybirds are aposematically coloured argues against mimicry being the single unifying factor in the evolution of this defence.

3

Body size

A central problem of evolutionary biology is to provide an explanation for the design of organisms (Stearns, 1982) and an important feature of design is size. Ladybirds vary in size both within and between species, and between sexes within a species. The adaptive significance of this variation is poorly understood.

INTRASPECIFIC PLASTICITY IN SIZE

Many species of ladybirds have been reared at different temperatures, different feeding rates and on different kinds of prey. That is, they have been reared in what are in effect different environments, which can be defined in terms of temperature, and the abundance and quality of food. Most of these studies make no attempt to record the changes in response of a particular genotype to the different conditions. It is likely that these studies were based on the offspring produced by relatively few beetles and most of the larvae survived to maturity. Therefore, the results represent mainly the degree to which the expression of a trait changes between environments, i.e. phenotypic plasticity.

The patterns of change in body weight are remarkably similar for species from a large number of genera of both aphidophagous and coccidophagous ladybirds. With increase in the daily food supply to the larvae there is both an increase in the weight and a decrease in the developmental time. That is, there is an inverse relationship between adult weight and developmental time (Figs. 3.1, 3.2). However, above a certain rate of food supply there is no further increase in weight or decrease in developmental time (Figs. 3.1, 3.3(A),(B)). When an excess of food is supplied each day but the larvae are reared at different temperatures, there is initially a

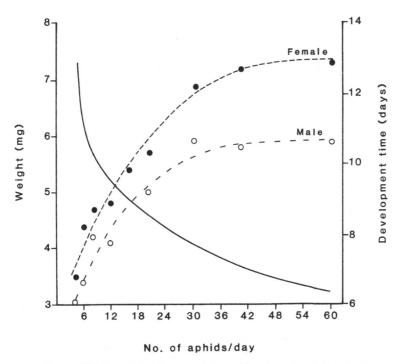

Fig. 3.1. Weight and developmental time of female and male *Propylea japonica* fed different numbers of aphids per day. (After Kawauchi, 1979.)

tendency for weight to increase with increasing temperature up to a maximum. The temperature at which the maximum weight is achieved varies between species (Fig. 3.4). However, generally there is a tendency for weight to increase at first and then decrease with increasing temperature (Fig. 3.3(D)).

By treating development and growth as separate processes it is possible to show that although ladybirds reared at high temperatures are small they nevertheless have a higher growth rate than the larger and similar-sized beetles reared at lower temperatures. Size is a consequence of the relative effect of food quantity and temperature on the growth and developmental rates. Both an increase in the availability of food and temperature result in an increase in the growth rate measured as increase in weight per unit weight per unit time (relative growth rate, RGR) and developmental rate ($1/D$, where D is the developmental time). However, increase in temperature and food supply affect these traits differently. For example, an increase in food supply results initially in an increase in both growth and developmental rates but the former increases faster than the latter (Fig. 3.3(A)). This results in the initial increase in weight observed in

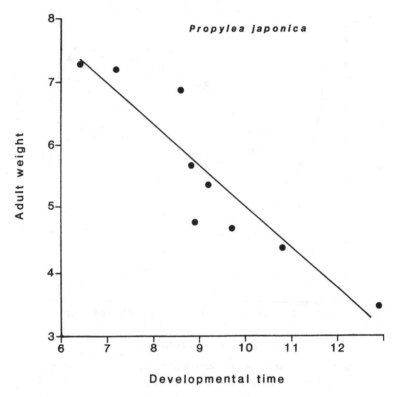

Fig. 3.2. Adult weight (in mg) in relation to developmental time of *Propylea japonica* supplied with different numbers of aphids each day. (After Kawauchi, 1979.)

the studies where larvae are reared on one of a range of food supplies from low to high. Above a certain level of food supply there is no further increase in weight, mainly because the larvae are satiated (Fig. 3.3(B)). Similarly, increase in the temperature at which larvae are reared is accompanied by an increase in both developmental and growth rates. However, the slopes of the two relationships differ and the lines intersect (Fig. 3.3(C)). As a consequence of the growth rate increasing faster than the developmental rate initially, weight first increases with increasing temperature but then decreases because further increase in temperature results in a disproportional decrease in the time it takes to reach maturity (Fig. 3.3(D)). What determines the form of the relationships between developmental rate and relative growth rate and temperature and food supply in different species of ladybird is unknown. However, the assumption is that these relationships have been shaped by selection and the

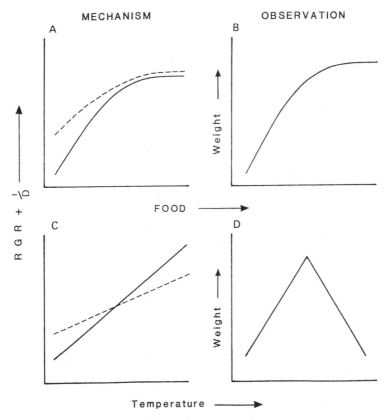

Fig. 3.3. The relationship between relative growth rate (RGR, ——) and developmental rate (1/D, ----), and food supply and temperature (mechanism) and the resultant increase in weight with increase in food supply or temperature (observation.)

resultant variation in adult size within a species is a consequence of maximizing the population growth rate, r_m (cf. Kindlmann & Dixon, 1992).

What life history traits are associated with size? Large females of *Coccinella septempunctata* have more ovarioles and are potentially capable of laying larger clusters of eggs than small females (Fig. 2.13(C)). Although there is variation in egg size this variation is not associated with adult size. It is likely that minimum egg size is constrained by the effect egg size has on both developmental time and the ability of first instar larvae to capture active prey. If relatively small at birth a larva takes longer to complete development and is inefficient at capturing prey (Stewart *et al.*, 1991b). If these constraints are limiting then it follows that in times of food stress ladybirds should vary the size of their clusters of eggs rather

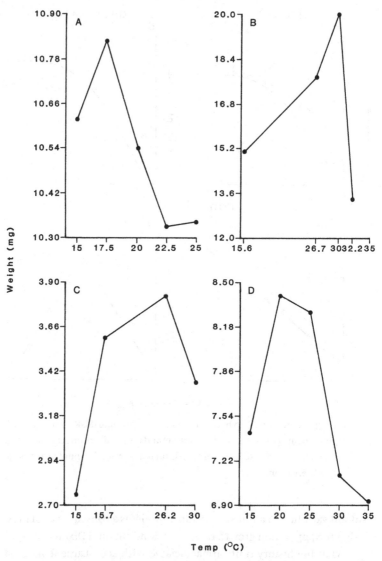

Fig. 3.4. The triangular relationships between adult weights and rearing temperature in (A) *Adalia bipunctata*, (B) *Hippodamia quinquesignata*, (C) *Scymnus frontalis* and (D) *Propylea quatuordecimpunctata*. (Data from: Kaddou, 1960; Mills, 1979; Naranjo *et al.*, 1990; Quilici, 1981.)

than the size of their eggs. This is what is observed when the rate of food supply to *C. septempunctata* is switched from a low to a high rate and vice versa (Fig. 2.12; Dixon & Guo, 1993). Large individuals of many species of insects have larger fat bodies than small individuals. Thus one would expect the larger individuals to survive hibernation better than small individuals. Although it is clear that the size of the fat reserves of lady-birds entering hibernation is important in determining their overwinter-ing survival the latter is not associated with size (p. 28). This may be because the size of the fat body may be more influenced by the availability of food prevailing after the beetle completes its development rather than during its development. Although the maximum size of a beetle's fat body is likely to be determined by the beetle's size it would be interesting to know whether large individuals are potentially capable of having pro-portionally larger or smaller fat reserves than small individuals.

Why ladybirds are the sizes they are is still an open question. Significant genetic variation in phenotypic plasticity for development rate and size exists in *Harmonia axyridis* (Grill *et al.*, 1997). That is, provid-ing selection pressure is strong enough then there could be changes in both developmental time and adult size. The only study on the effect of artificial selection on life history traits in ladybirds was done in an attempt to breed *Hippodamia convergens* with shorter developmental times in order to facilitate their production for use in biological control. After five generations of selection the developmental time was reduced by approximately 18%, and no associated effect on adult weight, fecundity or longevity was observed (Rodriquez-Saonia & Miller, 1995). This resulted in a 7% increase in the intrinsic rate of population increase. This relatively small change may be in part due to the fact that aphidophagous ladybirds develop at or close to the fastest rate sustainable in such insects (cf. p. 69). It may also account for the low genetic variation underlying both develop-mental time and body size recorded in *H. axyridis* (Grill *et al.*, 1997). The results of both these studies are likely to have been confounded by the dif-ference in the weight of the sexes (p. 42).

Interestingly, if one rears the larvae of *Adalia bipunctata* individu-ally, all supplied with excess food, then some develop quickly and are large and others develop slowly and are small, even after the confounding effect of the difference in the size of the sexes is removed (Fig. 3.5). The individuals that develop slowly also appear to eat slowly, as if inherently slow developers, and vice versa. The implication of this relationship is that selection for fast or slow development should result progressively in larger or smaller adults, respectively. In addition, if rates of development are positively related to rates of feeding then rearing offspring of beetles that have been selected for either fast or slow development, on a small quantity of food, should result in more of the offspring of the fast-developing strain dying than of the slow-developing strain. If shown to apply generally to ladybirds, then this could be a major component of the

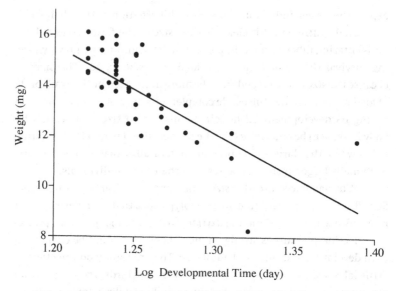

Fig. 3.5. The relationship between adult weight and developmental time for females of *Adalia bipunctata* reared under the same conditions of temperature and food supply. (Sato, personal communication.)

mechanism that underlies the slow–fast continuum in ladybird development, which is the subject of the next chapter.

SEX AND SIZE

Males are on average smaller than females in ladybirds (Fig. 3.1). As indicated above rearing ladybirds in different environments can result in adults of a wide range of weights. The data sets there are in which the weights of males and females are recorded separately indicate that the average weight of the male is always less than that of the female. The results for the two-spot ladybird, *Adalia bipunctata*, are illustrated in Fig. 3.6. Similarly, the relationship between the weight of the male (y) and that of the female (x) in *Propylea japonica* is:

$$\log y = -0.043 + 0.96 \log x \tag{3.1}$$

($r = 0.98$, $n = 16$; data from Kawauchi, 1979) and for *Leis dimidiata* is illustrated in Fig. 3.7. The relationship for *A. bipunctata* was obtained by rearing the ladybird at several different temperatures and feeding rates. That is, the relationship between male and female weights appears to hold over a wide range of conditions. That for *L. dimidiata* was obtained by

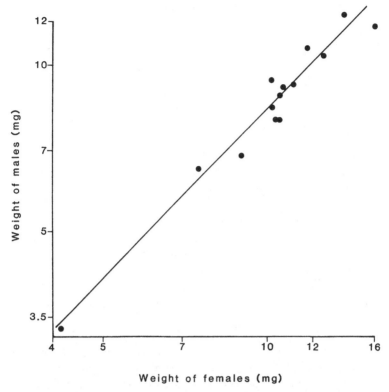

Fig. 3.6. The relationship between the logarithm of the weight of the males and that of the females of *Adalia bipunctata* reared at a range of temperatures and food supplies (log wt male = −0.07 + 0.97 log wt female, $r = 0.98$, $n = 14$). (After Mills, 1979; Yasuda & Dixon, 2000.)

rearing the ladybird on different species of aphid, which varied greatly in food quality. In these ladybirds the slopes of the relationships are not significantly different from 1.0, which indicates that male weight is a fixed proportion of female weight, and doubling the female weight is associated with a doubling of the weight of the male.

Data on the weights of the males and females of ladybirds of a wide range of size (0.8–66 mg) and feeding habits are available in the literature. Often there are several data sets for each species either published by different authors or as a result of the ladybird having been reared under different conditions by the same author. The above analyses indicate that it is permissible to average the data for each species as the proportional relationship of the weights of the sexes does not change with rearing conditions. The slope of the relationship between the logarithm of the weight of the male and that of the female for 29 species has a slope that is

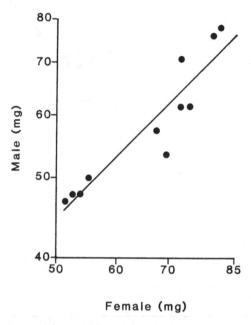

Fig. 3.7. The relationship between the logarithm of the weight of the males and that of the females of *Leis dimidiata* reared at the same temperature (25 °C) on different species of aphids (log wt male = −0.053 + 1.002 log wt female, $r = 0.93$, $n = 11$). (After Semyanov, 1996.)

significantly less than 1.0 ($P < 0.05$; Fig. 3.8). This indicates that interspecifically the weight of males as a proportion of that of females is less in large than in small species. This needs to be confirmed by measuring the weights of the sexes for more small and above all the largest species. If shown to be sound it introduces another dimension into sexual size dimorphism in ladybirds.

SEXUAL SIZE DIMORPHISM

In most poikilotherms sexual size dimorphism is female biased (Wicklund & Karlsson, 1988). While sexual selection may contribute to this bias the evolution and maintenance of sexual size dimorphism is a result of differences in the net selection pressures acting on the body size in both sexes (Arak, 1988). Several hypotheses have been proposed to account for the relatively small size of males in most poikilotherms. Firstly, it is a consequence of selection for rapid development and early

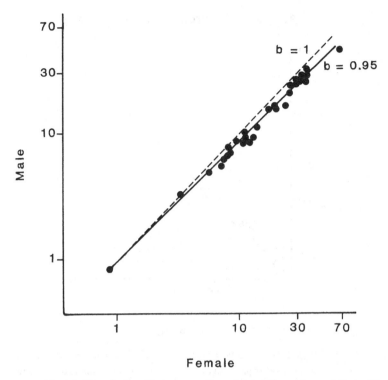

Fig. 3.8. The relationship between the weight of the males and that of the females for 29 species plotted on a logarithmic scale (log wt male $= -0.092 + 0.95$ log wt female, $r = 0.99$). (Data from: Agarwala, personal communication; Campbell *et al.*, 1980; El-Hariri, 1966; Ferran & Larroque, 1979; Ferran *et al.*, 1984; Formusoh & Wilde, 1993; Frazer & Gill, 1981; Hales, 1979; Hemptinne, J.-L. & Magro, A., personal communication; Hukusima & Ohwaki, 1972; Kaddou, 1960; Kaniku-Kiamfu *et al.*, 1992; Maelzer, 1978; Munyaneza & Obrycki, 1998; Naranjo *et al.*, 1990; Niijima *et al.*, 1986; Parry, 1980; Phoofolo & Obrycki, 1998; Quilici, 1981; Schanderl *et al.*, 1985; Smith, 1966a; Tadmor *et al.*, 1971)

maturation of males – *protandry* or *developmental constraint hypothesis* (Alexander *et al.*, 1979; Fairbairn, 1990). Secondly, males begin developing their gonads earlier in their development than females, and this has costs in terms of the growth rate that males can sustain. Thirdly, females should be bigger than males simply because a female's fecundity is directly related to her size. Fourthly, in mating systems dominated by scramble competition, where male reproductive success is a function of the rate of encountering females, small males may be favoured when food is limiting because they require lower absolute amounts of food, and can

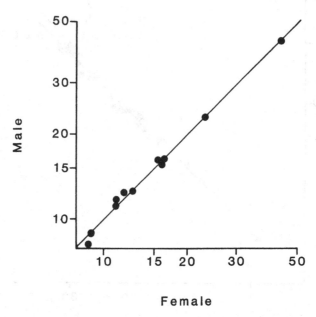

Propylea quatuordecimpunctata

Fig. 3.9. The relationship between developmental time of males and females of *Propylea quatuordecimpunctata* reared under a range of conditions of temperature and food supply, and plotted on a logarithmic scale. (After Quilici, 1981.)

therefore spend more time looking for females – *time and energy constraint hypothesis* (Ghiselin, 1974; Reiss, 1989).

Protandry

Surprisingly there are several data sets for the developmental times of males and females of species of ladybirds that feed on aphids, coccids, mites, psyllids and whitefly. Some of the species were also reared under a wide range of conditions of temperature, and of food quality and availability. In *Propylea quatuordecimpunctata*, the data indicate that although the developmental times are very dependent on the conditions of temperature and of food quality and supply, nevertheless the ratio of the developmental times of male and female larvae are not significantly different nor do they change with the conditions as the slope of the log/log relationship is not significantly different from 1.0 and the intercept is not significantly different from zero (Fig. 3.9). As the ratio of the developmental

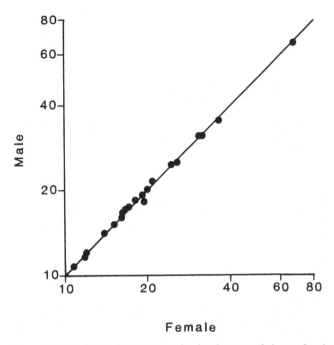

Fig. 3.10. The relationship between the developmental times of males and females of 16 species plotted on a logarithmic scale (cf. Table 3.1.)

times does not appear to be affected by the rearing conditions then the data sets for the individual species can be summarized by averaging the results for those species for which there is more than one record. This clearly indicates that in all the species studied the sexes show very similar developmental times (Fig. 3.10). Furthermore the results for 12 of these species were tested statistically and shown not to differ significantly (Table 3.1).

In addition to developmental time, growth rate and birth weight are also important in determining adult size (p. 27). As the developmental times of males and females are similar their growth rates or birth sizes may differ. Although there is variation in egg size within clusters of eggs laid by aphidophagous species of ladybirds the distribution of egg sizes tends to be unimodal rather than bimodal (Dixon & Guo, 1993), which argues against a gender difference in birth size. The idea, however, has only been specifically tested in A. bipunctata. In this species eggs destined to become males weigh 121 μg and females 120 μg (Yasuda & Dixon, 2000). Therefore the difference in the adult weights of males and females does not appear to be due to differences in their birth weights.

On the beetles reaching maturity several days elapse before they

Table 3.1. Average duration of development from egg to adult of female and male individuals of 17 species of ladybird

Species	Prey	Temperature, °C	Duration of development, days		Source
			Female	Male	
Adalia bipunctata	Aphids	20	23.7	23.2*	Yasuda & Dixon (2000)
		25	14.95	14.8*	Holloway et al. (1993)
Cleothora notata	Coccids		24.5	24.6*	Correjo et al. (1991)
Coccinella septempunctata	Aphids		31.03	31.13*	Yasuda & Dixon (2000)
Coccinella septempunctata brucki	Aphids		20.03	19.93*	Okamoto (1978)
Coleophora quadrivittata	Coccids		13.91	14.15	Chazeau (1981)
Harmonia axyridis	Aphids		25.8	25.1	Tan & Li (1932–3)
			16.35	16.58*	Okamoto (1978)
			13.00	13.06	
			13.46	13.47	
			14.04	14.54	
			14.50	14.60	
			14.56	15.08	
Hippodamia quinquesignata	Aphids	15.5	67.2	66.5	Kaddou (1960)
		27.0	15.0	15.0	
		30.0	10.6	10.6	
		32.0	11.9	11.9	
Menochilus sexmaculatus	Aphids		16.27	16.22*	Alikon & Yousuf (1986)
Nephaspis aculatus	Aleyrodids		19.4	18.3*	Liu et al. (1998)

Olla v-nigrum	Psyllids		16.69	16.93	Chazeau et al. (1991)
Propylea quatuordecimpunctata	Aphids		19.2	18.9	Quilici (1981)
			16.2	15.9*	Yasuda & Dixon (2000)
			16.0	16.1*	
Pharoscymnus numidicus	Coccids		31.53	31.07	Kehat (1967)
Pseudoscymnus tsugae	Coccids	20	29.8	29.8*	Cheah & McClure (1999)
		25	17.9	17.9*	
Rodolia iceryae	Coccids		36.1	36.4*	Kairo & Murphy (1995)
Stethorus madecassus	Mites		11.7	11.7*	Gutierrez & Chazeau (1972)
Stethorus picipes	Mites		16.86	17.24*	Tanigoshi & McMurtry (1977)
Stethorus punctillum	Mites		21.02	21.31*	Putman (1955)

Note:

*Were tested and shown not to differ significantly.

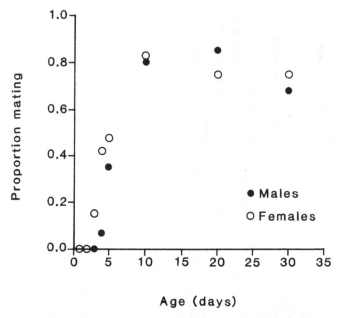

Fig. 3.11. The proportion of virgin males and females of *Adalia bipunctata* of different ages that copulated when presented with a sexually mature mate. (After Hemptinne *et al.*, 2000.)

become sexually mature and mate. The time to sexual maturity in adult ladybirds has only been recorded in detail for one species, *A. bipunctata*. In this species it appears that the females become sexually mature slightly earlier (5.3 days) than the males (6.7 days) when kept at 20 °C (Fig. 3.11; Hemptinne *et al.*, 2000a).

That is, the relatively small size of the males in ladybirds cannot be attributed to small size at hatching or protandry. However, this does not rule out the possibility that the similar overall developmental times and times of sexual maturity in males and females are achieved in a way that is more costly for males.

Gonadal constraint

The males of holometabolous insects, like Lepidoptera, are known to start developing their gonads before their female sibs (Reed & Beckage, 1997), and it is also known that this is associated with male dwarfism in these species. In addition parasitic castration is often associated with gigantism. Therefore, it is likely that the relative time of the onset of development of gonads in ladybirds might be a key factor deter-

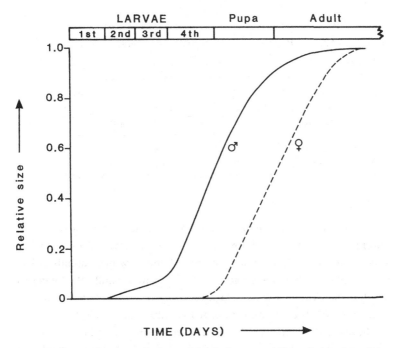

Fig. 3.12. The time of onset of development and the relative size of the gonads in the immature stages of *Adalia bipunctata*. (Agarwala, personal communication.)

mining the relative size of the sexes. What evidence is there for this hypothesis?

As in other holometabolous insects the gonads of *A. bipunctata* start developing earlier and achieve a greater biomass in the late larval stages of males than of females (Fig. 3.12). Associated with this earlier development of gonads there is a poorer conversion of aphid to ladybird biomass and a lower relative growth rate in male than female fourth instar larvae of *A. bipunctata* (Table 3.2). In both cases it is approximately 80% of that achieved by females and is similar in this respect to that of the difference in the adult weight of the sexes (Yasuda & Dixon, 2000). Although similar data need to be collected for other species the implication of the above is that gonads of males take longer to develop to maturity than those of females and therefore if the sexes are to reach sexual maturity synchronously then males need to start developing their gonads before the females. A possible consequence of this is that males are smaller than females. One possible mechanism is that the gonads in males start competing with the soma for resources earlier than in females and this reduces the growth potential of the soma of males. The cumulative affect

Table 3.2. *Food conversion and relative growth rates for the period from beginning of the fourth instar to adult emergence for female and male* Adalia bipunctata

	Food conversion	Relative growth rate
	$\dfrac{\text{Adult weight} - \text{Initial weight of fourth instar}}{\text{Weight of food consumed}}$	$\dfrac{\text{ln adult weight} - \text{ln initial weight of fourth instar}}{\text{Developmental time}}$
Female	0.075	0.070
Male	0.063	0.058

of this increases with each instar and as a consequence male larvae have a lower relative growth rate than female larvae. This is very similar to what is observed in the development of winged and unwinged aphids. It has been suggested that the winged aphids are smaller and take longer to develop than the unwinged individuals because their wing apparatus competes with their soma for limited resources (Dixon, 1998; Dixon & Kindlmann, 1999). That is, although there are only data for one species there is some evidence in support of the hypothesis that sexual size dimorphism in ladybirds is a consequence of the early development of gonads in males competing for a limiting resource.

Fecundity advantage

Darwin (1874) suggested that the general phenomenon of larger females was a consequence of the larger females being more fecund. The *fecundity advantage* model is the most common explanation of female-biased size dimorphism. However, it does not explain why females should be consistently larger than males.

In ladybirds large females have more ovarioles in their ovaries and when well fed lay more eggs per unit time than small females (Fig. 2.13; Dixon & Guo, 1993). However, a female's fecundity is possibly more limited by food availability than by her size.

Time and energy constraint

Male ladybirds do not appear to contest for mates and large males are no more effective than small ones at displacing the sperm of previous matings (Ueno, 1994). On emergence from hibernation proportionally more large than small non-melanic males of *Harmonia axyridis* are observed mating (Ueno *et al.*, 1998; Osawa & Nishida, 1992). However, this does not necessarily indicate that large males always have a mating advantage because it was not observed in melanic males, and the fat reserves of males emerging from hibernation could be size dependent. Under direct irradiation large ladybirds show a greater body temperature excess than small ones (Brakefield & Willmer, 1985; Stewart & Dixon, 1989), which would result in them being more active and more likely to mate when temperatures are low, and would cause a quicker depletion of fat reserves when temperatures are high. In addition the fact that males are consistently smaller than females argues against large males having an overall mating advantage. Therefore, there would appear to be no advantage in their being large and as in other arthropods that show a female-biased sexual size dimorphism the males appear to scramble for mates. Adult males of a wide range of species and prey types consistently consume less food and as a consequence spend less time feeding than females (Balduf, 1935; Obata & Johki, 1990; Table 3.3). This is to be expected as males do not have to develop large eggs; however, they spend less time resting, walk faster and fly more frequently than females (Obata & Johki, 1990; Tourniaire *et al.*, 2000a). Under laboratory conditions males generally have an adult life that is 15% shorter than that of females (Fig. 3.13; Table 3.4). This may also indicate that the high level of activity in early adult life in males, which has an advantage in terms of mating, has a cost in terms of reduced longevity.

In a world of limited, uniform-sized prey items, a large male would deplete the energy gained from its last meal faster than a small male as metabolic rate scales as mass to the power of 0.75 (Calder, 1984). Therefore, large males need to spend a greater proportion of their time foraging for food and a smaller proportion searching for females than small males. If true this might account for why the males of so many species are smaller than the females. This has been tested by rearing lady-birds on an abundant and a limited food supply, which resulted in the production of small and large males. These males were then either fed an excess of food or very few aphids per day. When large and small males were placed with females the number of small and large males that mated was dependent on their feeding regime. When well fed there was no significant difference in the mating performance of the males; however, when the food supply to the males was restricted a significantly greater proportion of the small males mated (Fig. 3.14; Yasuda & Dixon, 2000).

In summary, males of many species of a wide range of size, reared under a range of conditions of temperature and food supply, are

Table 3.3. *The food consumption of males expressed as a percentage of that of the females for 18 species of ladybird*

Species	Prey	Food consumption of male as % of female	Source
Adalia bipunctata	Aphids	58.5	Ellingsen (1960)
Cheilomenes lunata	Aphids	93.0	Ofuya (1995)
Chilocorus bipustulatus	Coccids	67.2	Yinon (1969)
Clitostethus arcuatus	Whitefly	45.1	Bathon & Pietrzik (1986)
Coccinella californica	Aphids	67.7	Frazer & Gill (1981)
Coccinella septempunctata	Aphids	68.6	Bodenheimer (1943)
		57.5	McLean (1980)
		88.7	Lucas et al. (1997b)
Coccinella transversalis	Aphids	87.0	Veeravel & Baskaran (1996)
Coelophora quadrivittata	Coccids	21.0	Chazeau (1981)
Cycloneda sanguinea	Aphids	77.4	Morales & Burandt (1985)
Harmonia axyridis	Aphids	40.0	Hukusima & Kamei (1970)
		56.0	Lucas et al. (1997a)
Hyperaspis raynevali	Coccids	72.0	Kanika-Kiamfu et al. (1992)
Menochilus sexmaculatus	Aphids	80	Veeravel & Baskaran (1996)
		51.9	Bose & Ray (1967)
		83.0	Gawande (1966)
		75.1	Varma et al. (1993)
		60.0	Bose & Ray (1967)
		44.0	
Pharoscymnus numidicus	Coccids	83.4	Kehat (1968)
Rhizobius lophantae	Coccids	33.0	Marin (1983)
Scymnus marginicollis	Aphids	52.0	Buntin & Tamaki (1980)
Stethorus madecassus	Mites	45.5	Chazeau (1974)
Stethorus gilvifrons	Mites	47.8	Kaylani (1967)
Stethorus punctillum	Mites	56.7	Putman (1955)

consistently smaller than females. This appears to be a consequence of males starting to develop their gonads earlier in larval development in order to synchronize their sexual maturity with that of the females. In addition small males require less food and therefore can spend more time looking for and mating with females.

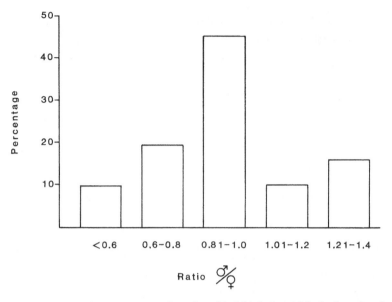

Fig. 3.13. The percentage of species of ladybirds in which the lengths of adult life of males divided by that of the females is <0.6, 0.6–0.8, 0.81–1.0, 1.01–1.2 and 1.21–1.4. (Data from: Agarwala & Choudhuri, 1995; Ahmad & Ghani, 1966; Bodenheimer, 1932; Campbell *et al.*, 1980; Ellingsen, 1960; Geyer, 1947*b*; Ginting *et al.*, 1992; Greathead & Pope, 1977; Hafez & El-Ziady, 1952; Hukusima & Kamei, 1970; Hukusima & Kouyama, 1974; Kaddou, 1960; Kapur, 1942; Kanika-Kiamfu *et al.*, 1992; Nsiama She *et al.*, 1984; Patnaik & Sahu, 1980; Rodriquez-Saona & Miller, 1995; Růžička *et al.*, 1981; Saha, 1987; Singh *et al.*, 1993; Varma *et al.*, 1993; Wille, 1926.)

BODY SIZE DISTRIBUTION

Theory

Although the predatory species of ladybird differ in size they are similar in shape (p. 10). The factors that determine the size of organisms are poorly understood (Haldane, 1927); however, their size is correlated with other attributes such as longevity, reproductive rate and resource use (Peters, 1983; Calder, 1984; Harvey & Pagel, 1991). Thus a knowledge of the distribution of body sizes within taxa might indicate differences in overall resource availability or differences in the way resources are partitioned among species (Pagel *et al.*, 1991).

The diversity of body sizes in organisms is thought to be a consequence of a random multiplication speciation process (Maurer *et al.*,

Table 3.4. *The number of species of ladybird in which the male has a shorter or longer adult life than the female*

	Length of adult life of male relative to that of the female		
	Shorter	Longer	χ^2
Number of species	14	2	4.01, $P < 0.05$

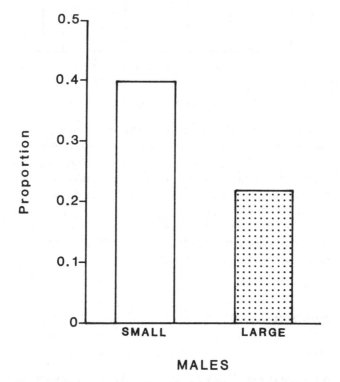

Fig. 3.14. The proportion of small and large males of *Adalia bipunctata* observed mating when food supply to the males is restricted. (After Yasuda & Dixon, 2000.)

1992). However, the size distributions of the species of a wide range of plants, herbivores and carnivores are similar in form, i.e. when size is plotted on a logarithmic scale the distribution is right-skewed (cf. Blackburn & Gaston, 1994). This skewness argues against accounting for the distribution purely in terms of speciation and extinction rates (Gould,

1988) and indicates that the distribution is shaped by directional changes (Maurer *et al.*, 1992).

Empiricists have tended to view size as optimum for the niche a species occupies (Stanley, 1973). In terms of functional biology the size of phytophagous and predatory organisms is thought to reflect the size of their host plants and prey respectively (e.g. Elton, 1927; Davidson, 1977; Kirk, 1991; Dixon, *et al.*, 1995; Kaspari, 1996). In contrast theorists have sought after a more general explanation, and have used the fractal structure of habitats or energetics to explain the evolution of body size distributions. The explanation based on fractal structure, i.e. there are more small niches suitable for small species than large niches for large species (e.g. Morse *et al.*, 1985), only offers a partial explanation as in none of the body-size distributions does the smallest class contain the most species. Brown *et al.* (1993) used a model based on energetics to account for the right-skewed size distribution observed in mammals. They argued that allometric constraints determine the efficiency with which resources are converted into offspring, which in turn determines the frequency of species of different body sizes. The peak in body size distribution is thought to coincide with the body size that is most efficient given the additional constraints of energy availability and interspecific competition. Recently Kozlowski & Weiner (1997) have proposed a model of life history evolution for animals that stop growing when they reach reproductive maturity, in which it is assumed that the rates of energy assimilation, respiration and mortality scale with body size according to simple power laws. Interestingly, although two of the six ecological parameters in this model are assumed to be normally distributed among species, the model predicts that the distribution of the body sizes should be skewed to the right even when the body size is logarithmically transformed. The other four ecological parameters were represented by means. If these are also varied then most of the predicted body size distributions are still skewed to the right but a few are now either not significantly skewed or even significantly skewed to the left (Kindlmann *et al.*, 1999). That is, this approach indicates that the distributions of body size can take several forms but are most likely to be logarithmically skewed to the right.

Empirical data

The ladybird faunas of the different regions of the world are not all equally well studied. The best-known biogeographic regions are the Palearctic, Nearctic, Ethiopian and Australian. In addition ladybird faunas are available for countries within the Palearctic, which make it possible to determine the effect of geographic scale on body size distributions. The data on the food of ladybirds also enable the hypothesis that ladybird size is correlated with that of their prey to be tested.

Size diversity

Regional differences

The total number of species of predatory ladybirds in the Palearctic, Nearctic, Ethiopian and Australian regions grouped according to size are given in Fig. 3.15. None of these distributions is normally or uniformly distributed. Those for the Nearctic and Australian regions are weakly skewed to the right. That is, the size distributions of predatory ladybirds differ from those of many organisms in not being significantly right-skewed except in the Nearctic. In the Nearctic there are proportionally more small species than in the Palearctic, and in the Australian region there are more species in the mid-size categories than in the Palearctic. In contrast, it is the numbers of species in the larger-size categories in the Ethiopian region that are greater than in the Palearctic region (Dixon & Hemptinne, 2000).

Difference within regions

The size diversity curves for six countries or areas within the Palearctic region (Fig. 3.16) are not normally distributed and in the case of four countries not uniformly distributed. More importantly, the distribution for Japan is significantly right-skewed and that for Central Europe significantly left-skewed. Relative to the Palearctic Japan has proportionally more small species. That is, the form of the body size distributions is not similar throughout the Palearctic (Dixon & Hemptinne, 2000).

Prey of ladybirds

A greater percentage of the small species, like *Stethorus* and *Scymnus*, feed on mites than of the large species; and similarly in the large species, such as *Ailocaria* and *Pseudosynonychia*, a greater percentage feed on chrysomelid and lepidopterous larvae (Fig. 3.17). If the prey of a ladybird is recorded, it is usually only the group to which the prey belongs: mites, aphids etc. Therefore, it is difficult to determine the actual size-range of prey in most cases. Another factor is the mobility of the prey. Aphids generally are more mobile than coccids, and large species of ladybird move faster than small species (Dixon & Stewart, 1991). Coccidophagous species of ladybird are on average smaller than aphidophagous species: data from

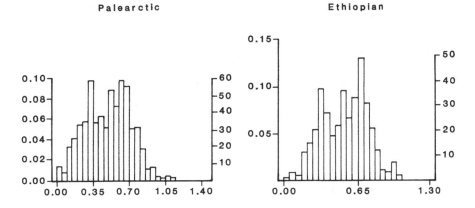

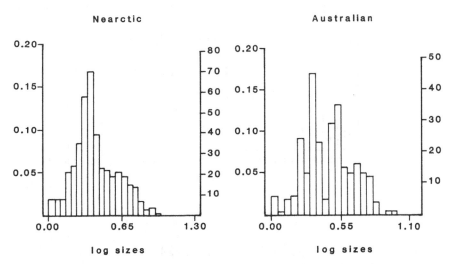

Fig. 3.15. The size diversity curves for the predaceous ladybirds in the Palearctic (609 spp.), Nearctic (424 spp.), Ethiopian (379 spp.) and Australian (269 spp.) biogeographic regions (left axis: relative frequencies, right axis: absolute frequencies, x axis: log body length in mm). (After Dixon & Hemptinne, 2000.) (Data from: Crotch, 1874; Gordon, 1976, 1985; Mader, 1926–37, 1941, 1950, 1954, 1955; Pope, 1988.)

Gordon (1976, 1985) give $\bar{x}_{coccid} = 2.6$ mm, $\bar{x}_{aphid} = 4.8$ mm, $t = 4.8$, df $= 37$, $P < 0.01$. Therefore, the increasing trend in the percentage of prey made up of aphids with increasing ladybird size (Fig. 3.17) is possibly a consequence of both mobility and the size of the prey.

The larvae of aphidophagous ladybirds usually have to pursue and subdue active prey whereas those of coccidophagous species are born

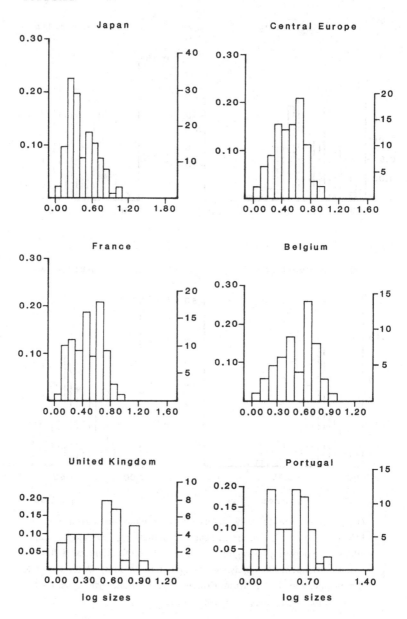

Fig. 3.16. The size diversity curves for the predaceous ladybirds in Belgium (54 spp.), Central Europe (91 spp.), France (92 spp.), Japan (146 spp.), Portugal (62 spp.) and United Kingdom (41 spp.) within the Palearctic region (left axis: relative frequencies, right axis: absolute frequencies, *x* axis: log body length in mm). (After Dixon & Hemptinne, 2000.) (Data from: Fürsch, 1967; van Goethem, 1975; Goureau, 1974; Pope, 1953; Portevin, 1931; Raimundo, 1992; Raimundo & Alves, 1986; Sasaji, 1971.)

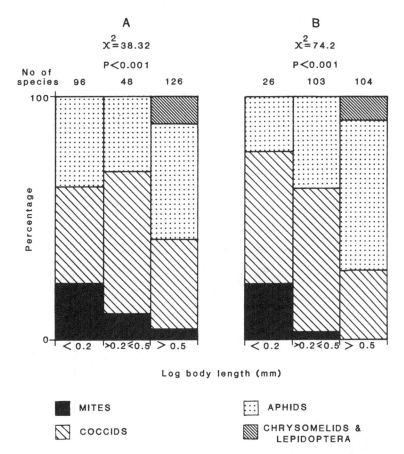

Fig. 3.17. The percentage of species of ladybirds of different body sizes specializing on mites, coccids, aphids, and chrysomelid and lepidopterous larvae. (After Dixon & Hemptinne, 2000.) (Data from: Collyer, 1953; Goureau, 1974; Hodek, 1973; Iablokoff-Khnzorian, 1982; Iwata, 1932*a,b*; Kamiya, 1961*a,b*, 1965, 1966; Mader, 1926–37, 1955; McKenzie, 1932; Sasaji, 1967*a,b*, 1968; Schilder & Schilder, 1928.)

surrounded by immobile prey. Both aphidophagous and coccidophagous ladybirds show the same fundamental relationship, with larger species laying larger eggs than small species (p. 26, Fig. 3.18). The slopes of the relationships for the aphidophagous and coccidophagous species plotted on a log/log scale are similar and both significantly less than 1.0. That is, both sets of data indicate that large species lay proportionally smaller eggs relative to their size than do the smaller species. However, as aphidophagous species are generally larger than coccidophagous species they lay larger eggs than the latter. That is, the empirical data support the

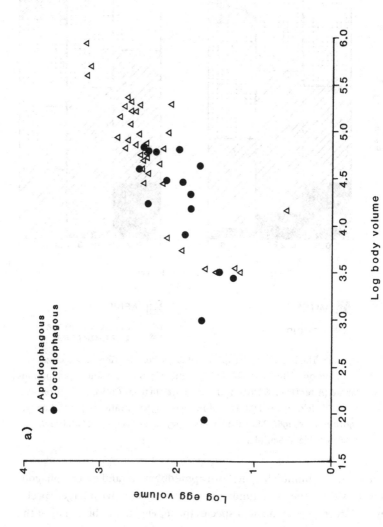

a)

△ Aphidophagous
● Coccidophagous

Log egg volume

Log body volume

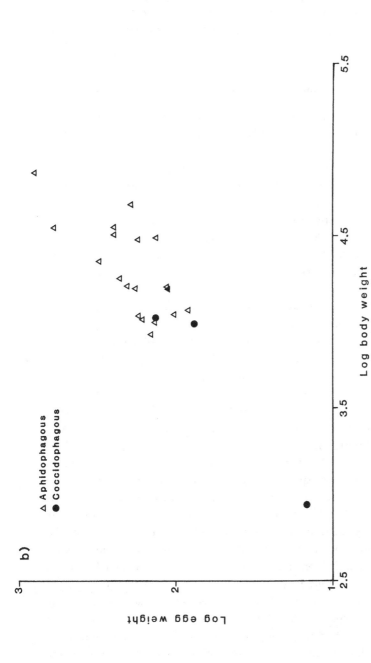

Fig. 3.18. The relationship between (a) the logarithm of the egg volume and that of adult volume for 52 species, and (b) between the logarithm of egg weight and adult weight for 22 species. (Volume = (length × breadth²) mm³ × 1000; weight = weight in mg × 1000; after Dixon & Hemptinne, 2000.) (Data from: Bodenheimer, 1932; Brettell, 1964; Chazeau, 1981; Davidson, 1923; Delucchi, 1954; Ginting *et al.*, 1992; Iwata, 1932*a,b*; Kesten, 1969; Longo & Benfatto, 1987; Maelzer, 1978; Mckenzie, 1932; Nsiama She *et al.*, 1984; Palmer, 1914; Quezada & DeBach, 1973; Simanton, 1916; Singh *et al.*, 1993; Tawfik, 1962; Tawfik & Nasr, 1973; Tranfaglia & Viggiani, 1972; Vandenburg, 1992, Vesey-Fitzgerald, 1940; Wille, 1926.)

contention that the difference in size between aphidophagous and cocci-dophagous species is adaptive because the offspring of aphidophagous species need to be large if they are to catch and subdue their more active prey.

What determines the shape of the size diversity curves?

As indicated above two models have dominated the recent litera-ture. The first based on the fractal structure of habitats (Morse *et al.*, 1985) and the second on energetics (Brown *et al.*, 1993). The relatively low numbers of small species in well-researched groups questions the credibility of the first explanation. The energetics-based model devel-oped for mammals, or a variant of it, is seen as offering the best poten-tial for a realistic description of the forces structuring body-size distributions (Blackburn & Gaston, 1994, 1996; Kozlowski & Weiner, 1997). These models predict not only mainly logarithmically right-skewed but also not skewed and skewed to the left size diversity curves (Kindlmann *et al.*, 1999). Few of the distributions observed in ladybirds are significantly right-skewed. The energetics-based model does not rule out the possibility that the body-size distributions of ladybirds are strongly dependent on that of the organisms on which they feed. The same is predicted by the Kozlowski & Weiner (1997) model if the assimil-atory, respiratory and mortality rates of the predator are dependent on the size of the prey.

The respiratory rates of similar-sized aphidophagous and cocci-dophagous ladybirds differ (p. 73). Although the efficiency with which ladybirds convert prey biomass into ladybird biomass is similar in these two groups of ladybirds the conversion occurs over a much longer time in the coccidophagous species (p. 69), which indicates that their assimila-tory rate is likely to be very much lower than that of aphidophagous species. That is, the little empirical evidence there is indicates that the assimilatory and respiratory rates of coccidophagous species of ladybird are considerably lower than those of similar-sized aphidophagous species. Similarly if one assumes that the abundance of aphidophagous and coccidophagous ladybirds changes little from year to year and accepts that the fecundity of the latter is lower than that of the former (p. 74) then the mortality rate of coccidophagous species is also likely to be less than that of aphidophagous species. If, as the above indicates, all the rates which according to Kozlowski & Weiner determine body size are lower in coccidophagous than in similar-sized aphidophagous species, then their model cannot account for body-size distributions in ladybirds.

Table 3.5. *Numbers of aphid and coccid species and aphidophagous and coccidophagous species of ladybirds in the Nearctic and Palearctic regions*

Region	Nearctic			Palearctic		
	Aphids	Coccids	χ^2	Aphids	Coccids	χ^2
Number of prey species[a]	1085	1183		2186	1938	
Number of predator species	162	186	0.16 (ns)	255	235	0.12 (ns)
Predator/prey ratio	0.15	0.16		0.12	0.1	

[a] Data from Eastop (1978) and Kozar & Walter (1985).

However, more data of this sort needs to be collected with the specific objective of testing the assumptions made by Kozlowski & Weiner.

In the absence of a general model there is a tendency to seek the directional process that has shaped the diversity curve of a specific group of organisms. What evidence there is tends to support the idea that each predator only eats prey of a limited range of body sizes (Elton, 1927; Sabelis, 1992). That is, morphological traits, including size, accurately reflect differences in resource utilization (Lack, 1947; Hutchinson, 1959; Peters, 1983; Calder, 1984). The size diversity curves for the ladybirds are not all logarithmically right-skewed and vary significantly in shape both within and between geographical regions (Figs. 3.15, 3.16). Compared to the Palearctic, the ladybird fauna of the Ethiopian region has a greater proportion of large species while the reverse is true of the Nearctic and Australian regions. How can these differences be explained? Approximately constant ratios of numbers of species of predators to number of species of prey have been recorded for several communities (Jeffries & Lawton, 1985; Pimm, 1991; Hall & Raffaelli, 1993; Begon *et al.*, 1996). That is, the number of prey species determines in some way the number of predator species (Pimm, 1991). Of the data sets, the Palearctic and Nearctic are the most extensively and reliably documented. In these two regions the ratios of the numbers of aphids to aphidophagous ladybirds and of coccids to coccidophagous ladybirds are similar (Table 3.5). As coccidophagous species of ladybird are usually smaller than aphidophagous species, the difference in the ratios of aphids to coccids in particular, and other groups of prey insects, could account for the difference in the body size distributions of the ladybirds in these two regions.

In addition to the correlation between type of prey and body size in ladybirds – with mites the prey of mainly the smallest species and chrysomelid and lepidopterous larvae the prey of mainly the largest species –

large species of ladybirds lay larger eggs than small species. Large eggs give rise to large larvae, which move faster and are capable of catching larger and more active prey than small larvae (Dixon, 1958; Glen, 1973; Sabelis, 1992). Life history theory predicts that organisms should optimize the trade-off between number and size of offspring. Thus, the positive interspecific relationship between adult size and offspring size argues for size being adaptive and associated with the need to pursue and subdue prey. That is, the body size distribution curves for predatory ladybirds appear to be dependent on the nature and relative abundance of their prey rather than their assimilatory, respiratory and mortality rates, which are possibly more associated with a particular way of life than body size.

In conclusion, ladybirds vary in size both within and between species. Within species, size is possibly the optimum for maximizing r_m and depends on the relative effects of food quality/quantity and temperature on the growth and developmental rates; and males are always smaller than females. Sexual size dimorphism appears to be a consequence of males starting to develop their gonads earlier in larval development than females, possibly in order to synchronize their sexual maturity with that of females. In addition small males require less food and therefore can spend more time looking for mates. That is, males and females should not be lumped together as adults, as is done in many studies, as their ways of life differ.

The regional differences in body size distributions do not appear to be determined by their assimilatory, respiratory and mortality rates but are due to large size in ladybirds being associated with large size and/or mobility of their prey, and vice versa. That is, it is likely the body size distributions in predatory ladybirds reflects that of their prey in the different biogeographic regions of the world.

4

Slow–fast continuum in life history parameters

INTRODUCTION

There is a linear relationship between the logarithms of generation times and body size for organisms ranging in size over almost eight orders of magnitude (Bonner, 1965). Among the placental mammals, such allometric relationships between life history parameters and body size have been reported for litter size, gestation length, duration of lactation, age at independence and sexual maturity, and maximum life span in captivity. Furthermore, placental mammals can be arranged in a 'slow–fast continuum', with large, long-lived, slowly reproducing species like elephants and whales at one end, and small species like rodents, with an opposite suite of traits, at the other end (cf. Harvey *et al.*, 1989; Promislow & Harvey, 1990). The general conclusion of the studies on mammals is that mortality, especially juvenile mortality, is the best predictor of variation in life history traits. After factoring out body size mammals with a high level of natural mortality tend to mature early and give birth to small offspring in large litters after a short gestation. Ladybirds also show similar marked differences in life history traits, but as with mammals a major confounding issue is body size. In addition, because ladybirds are poikilothermic, temperature is a further confounding factor.

Ecologists tend to view life histories as the result of natural selection acting on trade-offs imposed by the need to allocate finite resources between conflicting requirements, such as growth versus reproduction (Sibly & Calow, 1986). This view of the way in which life histories may have evolved is adopted in the following analysis of the life history parameters of ladybirds. The analysis is mainly restricted to aphidophagous and coccidophagous species because they are the best studied.

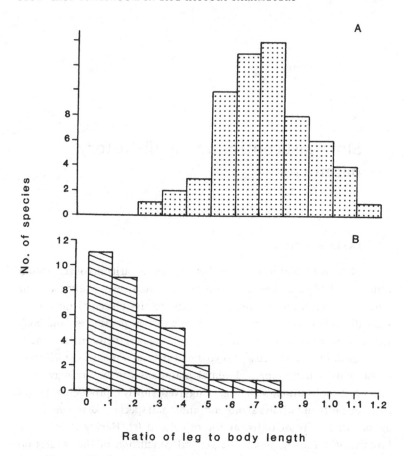

Fig. 4.1. The length of the legs relative to that of the body in (A) aphidophagous and (B) coccidophagous species of ladybird. (Relative leg length = total length of legs protruding beyond the edge of the body on one side / body length.) (Data from: Ahmad & Ghani, 1966; Bagal & Trehen, 1945; Bogdanova, 1956; Boving & Craighead, 1930; Brettell, 1964; Brown, 1972; Dixon, 1958; Eisner *et al.*, 1994; Geyer, 1947a; Ginting *et al.*, 1992; Gordon & Vandenburg, 1995; Hafez & El-Ziady, 1952; Houston, 1988; Kapur, 1942; Kesten, 1969; Palmer, 1914; Simanton, 1916; Tan, 1933–4; Toccafondi *et al.*, 1991; Vandenburg, 1992; Wille, 1926.)

SPEED OF MOVEMENT

Leg length of the larvae of aphidophagous and coccidophagous species differs greatly. Generally the relative leg length of aphidophagous species is greater then that of coccidophagous species (Fig. 4.1). The speed of movement (V) of the larvae of the aphidophagous *Calvia quatuordecim-guttata* is directly proportional to their leg length (L) ($\log V = 0.8 + 1.0 \log L$,

$r = 0.94$, $n = 80$; Vanhove, 1998). Thus within a species leg length and speed of movement are closely associated. Although several authors have recorded the speed of movement of the larvae and adults of ladybirds it is difficult to compare these results because they have been measured at different temperatures on beetles of different sizes on a variety of different substrates. However, the first instar larvae of the aphidophagous ladybird *Coccinella undecimpunctata* move approximately three times faster than the first instar larvae of the similar-sized coccidophagous ladybird *Cryptolaemus montrouzieri* (Magro, 1997; McLean, 1980). That is, in addition to their longer leg length there is evidence to indicate that the larvae of aphidophagous species generally move much faster than coccidophagous species. It is possible that the differences in speed of movement evolved first. The genus *Coelophora* is interesting in this respect as it contains one very specific coccid-feeder *C. quadrivittata*, which has a similar leg length but is less mobile than the two aphid-feeding species, *C. inaequalis* and *C. mulsanti* (Chazeau, 1981). Similarly, the larvae of *Coccinella magnifica*, which feeds on ant-attended aphids, have slightly shorter legs and move more slowly than the similar-sized *C. septempunctata* . This is again possibly associated with not having to pursue and subdue prey to the same extent as species feeding on unattended aphids and/or it is advantageous to have short appendages when attacked by ants. The latter would appear less likely as the larvae of *C. magnifica* are not attacked by ants (p. 105).

DEVELOPMENTAL TIME AND METABOLIC RATE

The relationships between developmental rate ($1/D$) and temperature for 29 species of aphidophagous and 19 species of coccidophagous ladybirds, each reared at a range of temperatures, indicate that aphidophagous species develop faster than coccidophagous species (Fig. 4.2). In terms of body size coccidophagous species are generally smaller than aphidophagous species (p. 59). That is, the smaller coccidophagous species spend longer in the immature stages than the larger aphidophagous species.

Interestingly the rates of development of each of these two groups of predators reflect those of their prey (Fig. 4.3), with aphids developing much faster than coccids. It is possible to select for faster development in ladybirds. Although the study of Rodriquez-Saonia & Miller (1995) did not reveal a cost in terms of reduced performance in another life history trait, theory predicts such a trade-off. The fittest individuals are likely to be those that most effectively harvest their prey (cf. Kindlmann & Dixon, 1993). Thus the upper points in the relationship for aphidophagous

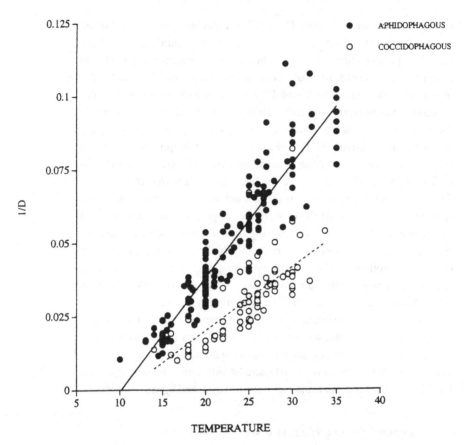

Fig. 4.2. The relationship between developmental rate (1/D) and temperature for aphidophagous and coccidophagous ladybirds. (D is the number of days from oviposition to adult emergence.) (After Dixon *et al.*, 1997.)

ladybirds in Fig. 4.2 are likely to be the fastest achievable by ladybirds. As the developmental rates of coccids are considerably slower than those of aphids the developmental rates of coccidophagous ladybirds have not been subjected to the same intense selection for faster development. On average, however, coccidophagous ladybirds have faster developmental rates than their prey.

As aphid and coccid eating tends to be confined to particular families and genera of ladybirds the difference in the rates of development recorded above may reflect phylogenetic constraint. However, there are genera like *Scymnus* 23% of the species of which feed exclusively on aphids and 62% on coccids (Hatch, 1961). An analysis of the data in the literature for five aphidophagous species and seven coccidophagous species indi-

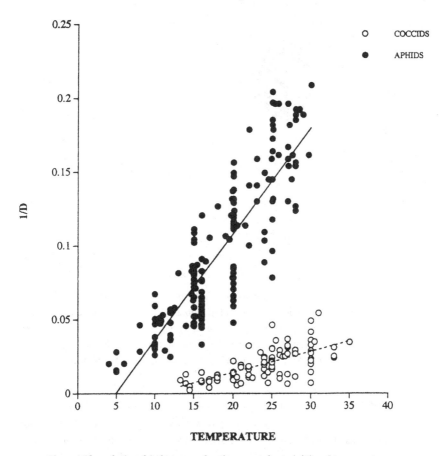

Fig. 4.3. The relationship between developmental rate (1/D) and temperature for aphids and coccids. (D is the number of days from birth or oviposition to the onset of reproduction.) (After Dixon *et al.*, 1997.)

cates that even within a genus the aphidophagous species develop faster than the coccidophagous species (Fig. 4.4; $F = 40.4$, df $= 37$, $P < 0.001$). This is strong evidence that it is the nature of the prey rather than phylogeny that determines the rate of development in ladybirds.

Further support for the idea that the speed of development of a predator reflects that of its prey comes from the few studies that have been done on phytophagous and other groups of predatory ladybirds. The developmental rates of phytophagous species, not surprisingly, are relatively slow and similar to that of coccidophagous species. Although the ladybirds that prey exclusively on other groups of arthropods are not as well studied as the aphidophagous and coccidophagous species, there are sufficient data in the literature to indicate their rates of development and

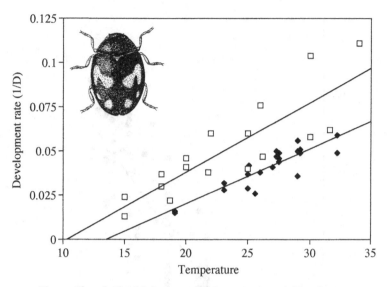

Fig. 4.4. The relationship between developmental rate (1/D) and temperature (°C) for five species of aphidophagous (□) and seven species of cocci-dophagous (◆) ladybirds of the genus *Scymnus*. (Data from: Buntin & Tamaki, 1980; de Fluiter, 1939; Mani & Thontadarya, 1987; Naranjo *et al.*, 1990; Rivnay & Perzelan, 1943; Tranfaglia & Viggiani, 1973; Zhao & Wang, 1987.)

those of their prey. Plotting the rates of development of the various groups of predatory ladybirds against those of their prey indicates that the rates do not increase proportionally. The curvilinear relationship indicates that there is an upper bound to the rate of development in lady-birds and that most groups of ladybirds develop more slowly than their prey (as discussed more fully in Chapter 9, p. 208). The exceptions are the coccidophagous species. That is, the rates of development of predatory ladybirds reflect those of their prey, but with a marked upper bound. The fact that acarophagous species of ladybird, like *Stethorus*, are considerably smaller than most aphidophagous species yet have a developmental rate similar to that of aphidophagous species argues against size being a deter-minant of rate of development.

The fact that coccidophagous ladybirds take considerably longer to develop than aphidophagous species has been attributed to the poor food quality of coccids compared with aphids. However, it is not only the devel-opment of the larval stages of coccidophagous species that is prolonged, but all stages (cf. Fig. 2.8). There are strong correlations between durations of the various stages in these two groups of ladybirds. As the egg and pupal stages contain all the nutrients necessary for development in a

Table 4.1. *The relative growth rates at 25 °C of six species of aphidophagous ladybirds and ten species of coccidophagous ladybirds calculated from data in the literature*

	Relative growth rate	Source
Aphidophagous species		
Adalia bipunctata	0.360	Mills (1979)
Coccinella septempunctata	0.520	Blackman (1967)
		Stewart *et al.* (1991*b*)
Harmonia axyridis	0.478	Schanderl *et al.* (1985)
Leis conformis	0.437	Maelzer (1978)
Menochilus sexmaculatus	0.580	Campbell *et al.* (1980)
Propylea quatuordecimpunctata	0.500	Quilici (1981)
Average	**0.479**	
Coccidophagous species		
Chilocorus bipustulatus	0.235	Hattingh & Samways (1994)
Chilocorus cacti	0.235	Hattingh & Samways (1994)
Chilocorus distigma	0.226	Hattingh & Samways (1994)
Chilocorus infernalis	0.270	Hattingh & Samways (1994)
Chilocorus nigritus	0.258	Hattingh & Samways (1994)
Cryptolaemus montrouzieri	0.210	Magro (1997)
Diomus hennesseyi	0.332	Kanika-Kiamfu *et al.* (1992)
Exochomus flaviventris	0.208	Kanika-Kiamfu *et al.* (1992)
Hyperaspis raynevali	0.283	Kanika-Kiamfu *et al.* (1992)
Nephus reunioni	0.309	Magro (1997)
Average	**0.257**	

high-quality form, it is unlikely that the overall speed of development is determined by the food quality of the prey. It is more likely that the speed of development has been determined by natural selection and is adaptive.

Both aphidophagous and coccidophagous ladybirds appear to be equally good at converting prey into predator biomass (see discussion in Chapter 9, p. 205). Their relative growth rates, however, are very different, with that of aphidophagous species being much greater than that of coccidophagous species (Table 4.1). Although the respiratory rates of very few ladybirds have been measured, that of the aphidophagous species *Menochilus sexmaculata* is over twice that of a similar-sized coccidophagous species, *Chilocorus bipustulatus* (Tadmor *et al.*, 1971). That is, it appears that certain ladybirds have been selected to develop rapidly and this is also reflected in their higher growth and respiratory rates.

FECUNDITY AND LONGEVITY

Ladybirds first grow and then on reaching the adult stage switch to reproduction. Life history theory predicts that reproductive activity should shorten adult life (Roff, 1992). However, there is little unequivocal support for this idea because of the confounding effect of an organism's size and condition on survival and mating success (Partridge & Harvey, 1985). Manipulative experiments with aphids, beetles and flies, however, clearly indicate that reproduction carries a cost in terms of future survival (Partridge & Farquhar, 1981; Roitberg, 1989; Tartar *et al.*, 1993; Dixon & Kundu, 1997; and p. 19). Although nothing is known about the length of life and fecundity of ladybirds in the field there have been many laboratory studies on their lifetime fecundity and longevity, although in any study only one of these parameters may be recorded. The data indicate their potential fecundity and longevity. Another complicating factor is that rearing conditions of temperature and food quality and/or quantity vary greatly between studies. For those species whose fecundity and longevity have been determined many times there is a clear but very variable trend – long life is associated with high fecundity (Fig. 4.5). Therefore, for those species that have been studied more than once the data used in the following analysis are those from the study that gave the longest longevity and/or highest fecundity.

The relationship between adult weight and fecundity (Fig. 4.6) indicates that the large species are more fecund than the smaller species and as aphidophagous species are generally larger than coccidophagous species that the former are generally more fecund than the latter. This tends to indicate that fecundity is simply a consequence of body size. Intraspecifically, however, longevity would appear to be an important determinant of fecundity. On average the adult longevity of coccidophagous ladybirds is significantly longer than that of aphidophagous species (Fig. 4.7). This suggests that the fecundity per day of adult life is less in coccidophagous than in aphidophagous species of ladybird. The interspecific plot of fecundity against longevity for both groups of ladybirds indicates, as was shown in the intraspecific relationship, a positive association between fecundity and longevity and more importantly shows that aphidophagous species produce more eggs per day than coccidophagous species (Fig. 4.8). The slopes of the relationships between fecundity (Fig. 4.6), egg weight (Fig. 3.17(b)) and adult weight (W) indicate that total potential fecundity in terms of weight of eggs produced per day is adult weight raised to the power of 1.4 ($W^{1.4}$). That is, not only are the generally larger aphidophagous species producing numerically more eggs, but

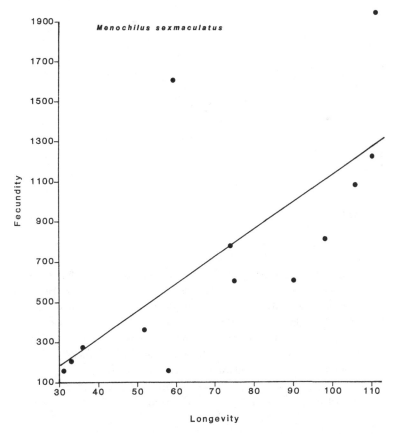

Fig. 4.5. The relationship between fecundity and longevity for *Menochilus sex-maculatus* based on data available in the literature. (Data from: Agarwala & Choudhuri, 1995; Babu & Ananthakrishnan, 1993; Campbell *et al.*, 1980; Gautam, 1990; Hukusima & Kouyama, 1974; Patnaik & Sahu, 1980; Saha, 1987; Varma *et al.*, 1993.)

these also possibly represent a greater investment in terms of adult weight than in the smaller coccidophagous species.

There have been several laboratory studies on the longevity of mated and virgin female ladybirds. These have all revealed that virgin females will lay eggs, but far fewer, and live longer than mated females. This also appears to apply to males (Table 4.2). Similarly, by manipulating the quality of the food supplied to adult *Adalia bipunctata* Kariluoto (1980) obtained markedly different fecundities and longevities, which reveal a directly proportional but inverse relationship between fecundity (*F*) and longevity (*L*) ($\log F = 3.83 - 0.94 \log L$, $r = 0.98$, $n = 4$). In *C. septempunctata* a

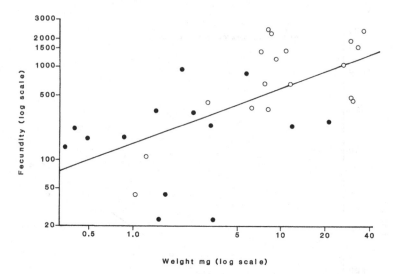

Fig. 4.6. The relationship between fecundity and adult weight for 18 species of aphidophagous (○) and 14 species of coccidophagous (●) ladybirds plotted on a logarithmic scale. (Data from: Agarwala *et al.*, 1988; Babu & Ananthakrishna, 1993; Blackman, 1965, 1967; Bogdanova, 1956; Booth *et al.*, 1995; Brettell, 1964; Campbell *et al.*, 1980; El-Hariri, 1966; Ferran *et al.*, 1984; de Fluiter, 1939; Geyer, 1947*b*; Gibson *et al.*, 1992; Greathead & Pope, 1977; Hafez & El-Ziady, 1952; Hukusima & Kamei, 1970; Kanika-Kiamfu *et al.*, 1992; Kapur, 1942; Kawauchi, 1981, 1985; Magro, 1997; Patnaik & Sahu, 1980; Ruzicka *et al.*, 1981; Singh *et al.*, 1993; Sundby, 1968; Wille, 1926; Wright & Laing, 1978.)

high proportion of the females do not reproduce continuously but intermittently. The females that show discontinuous reproduction have a lower fecundity and longer adult life than those that reproduce continuously (Růžička *et al.*, 1981). That is, as in other insects intraspecifically reproduction does appear to be costly in terms of length of adult life. Further evidence that reproduction is costly in terms of longevity comes from the observation that the fecundity of individuals of *C. septempunctata* which hibernated twice was lower during the first summer than those that only hibernated once (Sundby, 1968).

TRADE-OFF

Life history theory leads one to expect a trade-off in ladybird life history traits, with the 'fast' and 'slow' species at the two extremes of such a trade-off. There does appear to be a trade-off between adult longevity

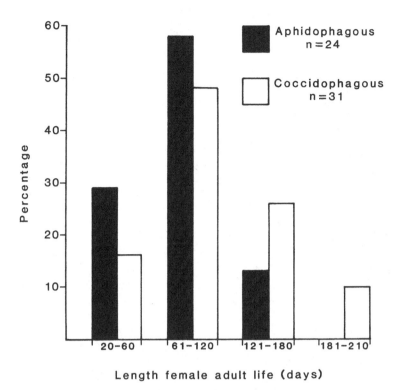

Fig. 4.7. The percentage of aphidophagous and coccidophagous ladybirds the females of which have been recorded as living for 20–60, 61–120, 121–180 and 181–210 days respectively.

and fecundity. Evidence for this is presented in Figs. 4.7 and 4.8. The average longevity of coccidophagous ladybirds is 107 days and of aphidophagous ladybirds 84 days (Fig. 4.7) and the associated fecundities are 295 and 646 eggs respectively (Fig. 4.8). More detailed studies on a few species tend to support this in indicating that the relative size of the gonads in coccidophagous species is half that in aphidophagous species (p. 27). This illustrates the very clear difference between aphidophagous and coccidophagous ladybirds: associated with the very rapid development of aphidophagous ladybirds is a short adult life and high fecundity, and with the slow development of coccidophagous ladybirds a long adult life and low fecundity (Fig. 4.9). However, adult longevity is closely associated with developmental time. If longevity is converted to rate of ageing, then each group of ladybirds appears to have a suite of life history traits, which are either all fast or all slow (Table 4.3). If this is the case then the inverse relationship between adult longevity and fecundity is not a

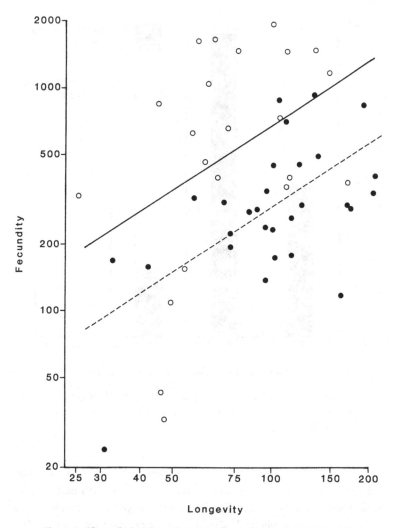

Fig. 4.8. The relationships between fecundity and longevity for aphi-dophagous (○) and coccidophagous (●) species of ladybird plotted on a log-arithmic scale.

consequence of a trade-off but of the beetles' pace of life. The implication of this is that it is not possible to have a short larval development coupled with a long adult life and low fecundity, or a long larval development coupled with a short adult life and high fecundity. That is, suites of traits are linked and selection acts on a set of traits rather than the individual traits. Although an attractive concept it should be treated with caution and subjected to a more rigorous analysis. It is possible the use of averages

Table 4.2. *The length of adult life of mated and virgin females and males of two species of ladybird*

| Species | Sex | Length of adult life | | Correlation with weight | Source |
		Mated	Virgin		
Adalia bipunctata	Female	77.4	109.6	No	Saâdaoui (1996)
	Male	90.0	103.1	No	
Scymnus interruptus	Female	86.1	147.7	(not tested)	Tawfik & Nasr
	Male	65.1	118.9	(not tested)	(1973)

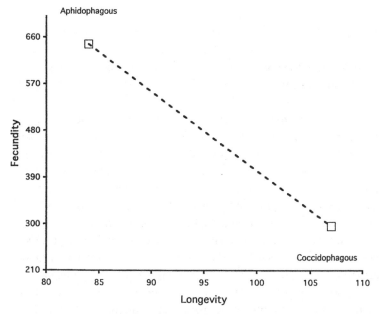

Fig. 4.9. The 'trade-off' between fecundity and longevity indicated by the average values of these parameters for aphidophagous and coccidophagous ladybirds.

could have obscured important variation at the species level. That is, the devil could be in the detail. There is now an urgent need for detailed studies on similar-sized species of aphidophagous and coccidophagous ladybirds specifically aimed at testing this idea.

The longer adult life of coccidophagous ladybirds indicated by the results in the literature needs to be confirmed by more detailed and specific experimental studies. If substantiated the adaptive significance

Table 4.3. *Summary of the differences in the life history parameters of aphidophagous and coccidophagous ladybirds*

	Aphidophagous	Differences	Coccidophagous
Rate of development		>	
Relative growth rate		>	
Metabolism		>	
Speed of movement		>	
Reproductive investment		>	
Reproductive rate		>	
Rate of ageing		>	

of long adult life in coccidophagous species becomes an important issue. The coccid prey of these ladybirds generally appears to be very much less abundant than the aphid prey of aphidophagous species. If this is true then a long adult life could be advantageous in that it gives the beetles more time to locate prey.

Thus, there appear to be suites of life history parameters that are always associated. Some support for this comes from the seminal paper of Rodriquez-Saona & Miller (1995). After selecting *Hippodamia convergens* for fast development they observed that fast-developing larvae consumed more prey per unit time and developed into bigger adults than more slowly developing individuals. More studies of this kind are needed. However, selecting aphidophagous species for fast development is only likely to result in small changes in life history traits as it is likely these species have been strongly selected for fast development (p. 69). Current attempts to select for faster development in coccidophagous and slower development in aphidophagous ladybirds should reveal the extent to which these life history parameters are linked and provide an experimental test of the trade-off between longevity and fecundity in ladybirds.

In conclusion, the coccidophagous species appear to have a slow and the aphidophagous species a fast pace of life, which affects all their life history characteristics. This very strikingly reflects the speed of life of their respective prey. Aphids being parthenogenetic and telescoping generations have achieved prodigious rates of increase compared with coccids. That is, the slow–fast continuum in ladybirds reflects the pace of life of their prey. Although there does appear to be a trade-off between fecundity and longevity this may not be a conse-quence of these two groups of ladybirds allocating different proportions of a finite resource to fecundity and longevity. All the evidence seems to indicate that the life history traits of these ladybirds are all either fast or slow. If substantiated this could greatly simplify the modelling of predator–prey interactions, but would appear to pose a problem for the idea of resource limitation being the major factor shaping life history traits.

5

Foraging behaviour

INTRODUCTION

The central biological problem is the design for survival (Williams, 1966) and a major factor in this is foraging behaviour. Organisms spend a lot of time either searching for or exploiting resources. The immature stages of ladybirds, like those of other insect predators, have very limited powers of dispersal compared to the adults, which search for and select the patches of prey that their larvae exploit. Most studies on foraging behaviour in ladybirds have been on the larvae, possibly for a very good pragmatic reason – it is the easiest stage to study. However, theory indicates that it is an adult's ability to select suitable patches for oviposition that is likely to be most important in determining its fitness (Kindlmann & Dixon, 1993).

FUNCTIONAL RESPONSE

There has been a tendency to concentrate on the functional response shown by ladybirds to prey abundance mainly because of its perceived importance in population dynamics (p. 157). For example, it is thought that a more efficient predator attacks a larger proportion of the prey over a given period of time than does a less efficient one and the attack rate (search rate), which can be derived from the functional response, is a good measure of this (Hassell, 1992). This has resulted in the proposal that functional responses should form the basis for comparative studies of ladybirds (Houck & Strauss, 1985). Functional responses have been incorporated into models used to account for the mortality of aphids in the field (Frazer & Gilbert, 1976; Baumgaertner et al., 1981). This approach has culminated in a series of multi-trophic models in which the supposed critical parameters of successful natural enemies, area of discovery and reproductive capacity, are replaced by a new approach in which physiology drives the functional response and how species and individuals allocate their resources, and an organism's behavioural responses to varying weather conditions (Gutierrez *et al.*, 1984, 1994). These models depend on obtaining an accurate estimate of the number of adults present in a patch or crop. Once this is known then the functional response(s) determines the number of prey caught per unit time and this combined with how the adults allocate their resources determines oviposition, and the beetles continue to oviposit so long as there are sufficient aphids to sustain egg production (Gutierrez *et al.*, 1984). This approach tends to regard ladybirds mainly as a means of reducing the abundance of aphids, i.e. as potential biological control agents, rather than organisms attempting to maximize their fitness.

One consistent factor emerges from the enormous literature on functional responses: experimental design often determines the shape of the response. Confinement to a specific area for a specific time is obviously unnatural and there is a danger that the behaviour is uncharacteristic or is assigned the wrong motivation. In most cases predators' behavioural responses are adapted to exploiting non-random distributions of prey, and they aggregate where prey is abundant. Adult ladybirds seek out high-density patches of prey for breeding. Thus, the 'functional response' of ladybirds in patches containing low numbers of prey would appear to be irrelevant as they tend to avoid such patches and aggregate in those that contain high numbers of prey. This is particularly true for larvae, which are confined to the patch of prey selected by their mother.

Foraging behaviour will only be better understood if it is viewed more as a means of maximizing ladybird fitness than of reducing the abundance of their prey. In this chapter the foraging behaviour of adults and larvae will be considered separately in order to understand better its adaptive significance and to establish the different roles and behaviours of these two stages. However, before doing this it is necessary to consider what is likely to be involved in prey recognition and how that may have affected prey specificity in ladybirds.

PREY RECOGNITION

Virtually nothing is known about this aspect of ladybird ecology yet it is central to discussions of diet breadth, cannibalism and intraguild predation. Adults in selecting an habitat and then a patch of prey play a major role in determining the range of potential prey available to their larvae. As patches of prey are attractive to several species of predator it is likely that each patch could contain more than one species of aphid and several species of predator. Therefore, ladybirds are likely to encounter a wide range of insects. This is particularly true of larvae. How do they decide which to attack and eat?

Relative risk as a determinant of diet breadth

Current studies on cannibalism and intraguild predation in lady-birds are beginning to indicate that the most important sensory input for foraging ladybird larvae is the chemical nature of the cuticular waxes (alkanes) of the insects they encounter (p. 87). It is likely that the waxes of the species of a particular group of prey are similar and differ from those of other groups of prey insects, similarly for predators. That is, each group of prey or predators has a particular chemical image. Support for this comes from the observation that the aphidiid parasitoid *Lysiphlebus cardui* mimics the cuticular hydrocarbon profile of its host the black-bean aphid, *Aphis fabae cirsiiacanthoidis*, and thereby is able to forage in ant-attended aphid colonies without being attacked by ants. There is great qualitative resemblance in the hydrocarbon profiles (n-alkanes, non-methyl, dime-thyl and trimethyl alkanes) of the parasitoid and the aphid. Conversely, another parasitoid (*Trioxys angelicae*) of the black-bean aphid, which has a different alkane profile from the aphid, is vigorously attacked and even killed when encountered by ants (Liepert & Dettner, 1996). All stages of development of ladybirds also appear to be similarly coated with species-specific alkanes. In this case to attack something with the same alkane profile as yourself could be risky as instead of being the cannibal you could become the victim. If the alkane profile is that of another species of predator then the risk could be greater because even if successful the victor may be poisoned (p. 184; Fig. 5.1) (Hemptinne *et al.*, 2000b). Further support for this comes from the observation that one of the preferred aphid prey of the ladybirds *Adalia decempunctata* and *Calvia decemguttata* can be made unacceptable by painting it with the cuticular alkanes from the surface of the eggs of *C. quatuordecimguttata* of which the contents are toxic to these ladybirds (unpublished results). That is, the surface alkanes appear to flag the nature of the organism encountered.

Pursuing, subduing and eating prey is not without risk. For example, if a ladybird is small relative to an aphid then it is at risk of being kicked off a plant, which could result in its death, or aphids may daub so much siphuncular wax onto a larva that it is incapable of moving and starves to death (Figs. 5.2, 5.3). Other aphids like *Hyalopterus pruni* and *Megoura viciae* are distasteful and poisonous, respectively, to some species

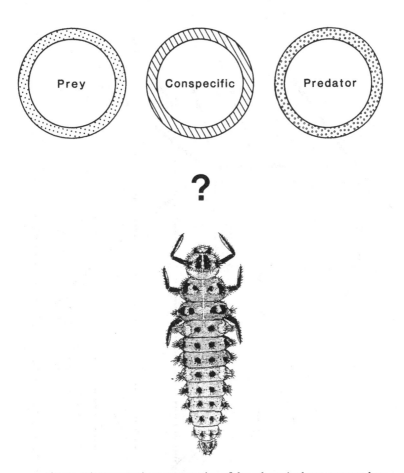

Fig. 5.1. Diagrammatic representation of the role cuticular waxes may have in prey selection in ladybirds.

of ladybirds (Dixon, 1958). In some cases the species of host plant of an aphid can determine its suitability as prey. For example, the larvae of *Adonis variegata* that feed on *Aphis nerii* living on *Cionura erecta* develop into apterous or brachypterous adults, whereas those that feed on the same aphid living on *Cynonchum acutum* or *Nerium oleander* develop normally (Pasteels, 1978). Larvae of *Coccinella septempunctata* fed on *Macrosiphum albifrons* from bitter lupins, such as *Lupinus albus, L. angustifolius* and *L. mutabilis*, which contain high concentrations of the alkaloid lupanine and 13-OH-lupanine, showed abnormal development or died, whereas those fed aphids from sweet lupins developed normally (Emrich, 1991). Ladybird larvae can learn to avoid unpleasant-tasting aphids, as larvae of

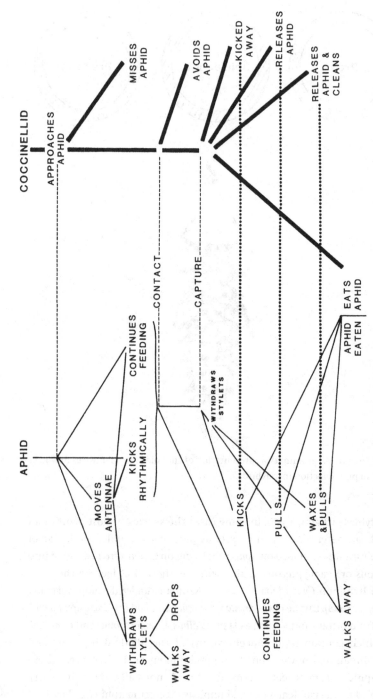

Fig. 5.2. Diagrammatic representation of the interactions that follow an encounter between a larva of *Adalia decempunctata* and a nettle aphid, *Microlophium evansi*. (After Dixon, 1958.)

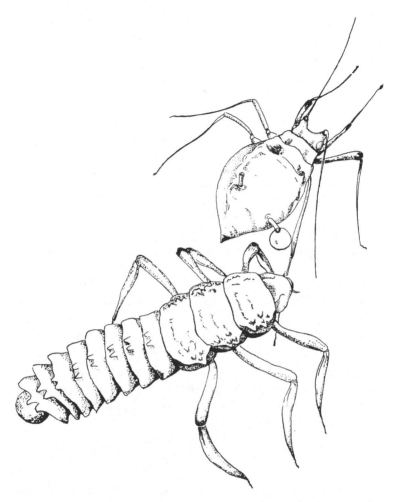

Fig. 5.3. An aphid attempting to escape from a ladybird larva that has seized one of its hind legs. The aphid attempts to pull its leg free, at the same time daubing a droplet of siphuncular wax on the head of the ladybird larva.

Adalia decempunctata that catch and feed on *H. pruni* will after a short while release the aphid. On encountering another individual of this species of aphid they will seize it but release it before starting to feed. That is, they appear to be able to learn to avoid feeding on certain aphids, which reduces the risk of poisoning. Thus, overall, attacking both aphids and predators is risky; however, the risk associated with intraguild predation is greater than that associated with cannibalism, which is more risky than attacking prey (Fig. 5.4).

In summary, attacking aphids is not without risk but the risk is

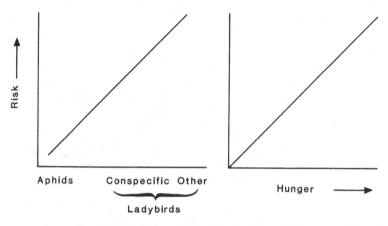

Fig. 5.4. The risk of death associated with attacking aphids, conspecific and heterospecific ladybirds, and from starvation.

likely to be less than that associated with cannibalism and above all intra-guild predation. However, hungry larvae are also at risk from starvation (Fig. 5.4). Therefore, the decision to attack is likely to be dependent on the relative risks associated with attacking and ingesting particular prey, and of starvation. The prediction is that the probability of attack increases with hunger, and that the hunger threshold for attacking aphids is lower than for attacking conspecifics, which is lower than for attacking intra-guild predators (Fig. 5.5). It is likely that cuticular alkanes play an important role in this decision.

PREY SPECIFICITY

Unlike plant tissues, which are often deficient in essential nutrients, animal tissues are more likely to supply predators with nutritionally balanced diets. Therefore other features such as the cost of capture and the toxin content are likely to be important in determining the choice of prey (Malcolm, 1992). Initially, whether a particular prey organism was regarded as suitable was based on whether the ladybird had been observed feeding on it. As this confused 'acceptability' and 'suitability' the concepts of 'essential food: that which can sustain larval development and oviposition', and 'alternative food: that which serves only as a source of energy to prolong survival', was developed (Hodek, 1959, 1962). In addition, there are the rejected or toxic prey that are not eaten (Hodek & Honěk, 1996). Food specificity was an important issue in ladybird ecology in the 1950–60s, with 25% of the text of Hodek's (1967) review of the predaceous species and a whole section of the first aphidophaga

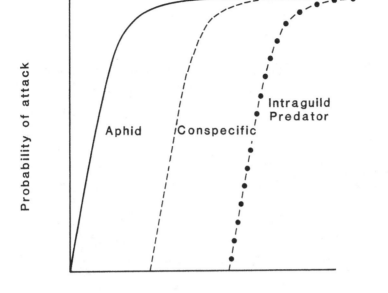

Hunger

Fig. 5.5. The likely relationship between hunger and the probability that a ladybird will attack an aphid, a conspecific or an intraguild predator.

meeting in 1965 devoted to this subject. It is still a major issue (Hodek & Honěk, 1996) with the emphasis mainly on pattern rather than process. Recently Malcolm (1992) suggested that when referring to the prey of a generalist predator the term 'included' be used for those prey that can be exploited successfully, 'peripheral' for those that are eaten but result in a significant decline in fitness, and 'excluded' for those that cannot be exploited because the predator is killed by them. This terminology is similar to that proposed by Hodek, with 'included' equivalent to 'essential', 'peripheral' to 'alternative' and 'excluded' to 'toxic'.

Thompson (1951) was the first to suggest that many ladybirds are not generalists but have specific food requirements. Certainly some of the coccidophagous species are more prey specific than aphidophagous ladybirds (Jalali & Singh, 1989; Kairo & Murphy, 1995; Strand & Obrycki, 1996). This may be a consequence of the relatively immobile coccids investing more in defence, in the form of a tough covering scale or toxins, than aphids, which often appear to depend more on their mobility to avoid capture (Dixon, 1958). That is, the greater host specificity of coccidophagous species may be a response to more strongly defended prey as has been suggested for chrysopids (Bristow, 1988).

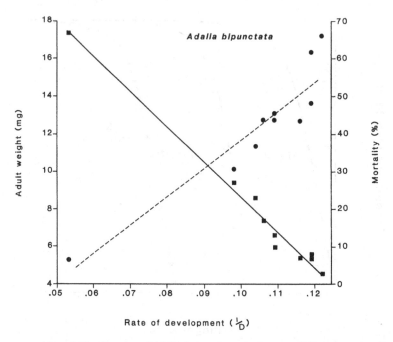

Fig. 5.6. The fitnesses of *Adalia bipunctata* reared on ten different species of aphid. When reared on high-quality prey developmental times are short, mortality is low (■——■) and adults are large (●----●), and vice versa. (After Kalushkov, 1998.)

The larvae of many aphidophagous species are frequently recorded eating a wide range of aphids. As a source of food aphids are not all equally suitable, as the larval developmental times and survival, and the adult weights when reared on these aphids vary greatly (Fig. 5.6; Kalushkov, 1998). That is, exploiting certain aphids would appear to result in a significant decline in fitness. In addition, although many aphids are suitable prey once caught, they are difficult to catch (Fig. 5.2; Dixon, 1958); or structures on the surface of plants, such as wax, glands and hairs, impede the movement of ladybirds, especially larvae, and make it difficult for them to catch prey (p. 123). This aspect of the interaction between ladybirds and their prey tends to be overlooked because laboratory assessments of prey quality are usually done in small uniform arenas in which prey are unable to escape.

Mills (1979) recorded the distribution of two-spot ladybird (*Adalia bipunctata*) eggs on deciduous trees infested with different population densities of aphids during the period of peak oviposition of this ladybird. It responds to an increase in the biomass of each aphid by ovipositing an

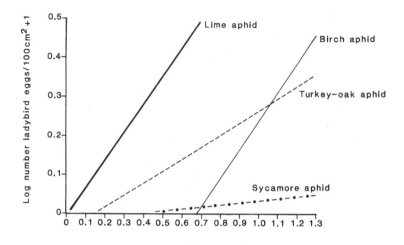

Fig. 5.7. The numerical responses, in terms of eggs laid per unit area of foliage, of *Adalia bipunctata* to increasing biomass per unit area of four species of tree-living aphids. (After Mills, 1979.)

increased number of eggs at a decreasing rate. The exploitation or numerical response to each species of aphid varies in the density at which oviposition begins and the rate at which oviposition declines with increasing biomass of aphids per unit area (Fig. 5.7). By using biomass one corrects for the differences in size of the different aphids. The responses to the lime aphid, *Eucallipterus tiliae*, and the birch aphid, *Euceraphis punctipennis*, are similar in slope but different in intercept. The birch aphid is more difficult to catch than the lime aphid, which possibly accounts for the difference in the intercepts of the relationships for these two species. The slopes of the numerical responses in all cases are significantly less than 1.0, which indicates that the rate of oviposition declines with increasing aphid infestation. This has been interpreted as a consequence of the poor ability of ladybirds in locating the few trees that support high aphid abundance (Mills, 1979). In years when aphids are very abundant the numerical response is not linear but plateaus and remains constant at the higher aphid population densities. The constant response at high aphid population densities is attributed to satiation (Mills, 1982b). The fact that the slopes of the responses to the oak and sycamore aphids differ from those for lime and birch aphids tends to argue that the attractiveness of aphids to ladybirds is not only determined by their abundance, escape ability, nutritional quality, size and host plant structure. If it were one would expect all the relationships to have the same slope. Individuals

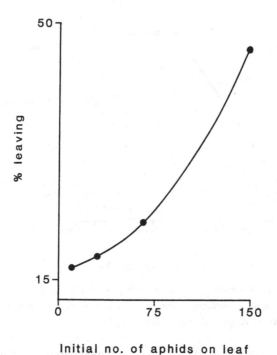

Fig. 5.8. The proportion of sycamore aphids leaving a leaf after a shadow has passed over the leaf relative to the original number of aphids on a leaf. (After Dixon & McKay, 1970.)

of some species of aphids, such as the sycamore aphid, *Drepanosiphum platanoidis*, react to one another's movements and as a consequence are more responsive to a disturbance caused by the presence of natural enemies when abundant than when scarce (Fig. 5.8; Dixon & McKay, 1970). When captured by a predator, aphids secrete an alarm pheromone, which causes adjacent aphids to scatter (Kislow & Edwards, 1972). It is also likely that this scattering response is more marked when aphids are abundant. That is, the abundance of an aphid can affect its relative availability to a predator.

A species of aphid that is suitable prey for one species of ladybird may be distasteful or toxic for other species of ladybird. Some ladybirds appear to attack many species of prey, whereas others have been recorded attacking very few species. Ladybirds show habitat specialization and whether they attack few or many species of prey may be related in part to the number of species of prey they regularly encounter in their respective habitats. In addition to the factors outlined above, whether a particular species of prey is regularly attended by ants is also likely to determine whether it is a suitable prey for a particular ladybird. For example, the larvae of *Platynaspis luteorubra* and *Coccinella magnifica* are only found feeding on ant-attended aphids. The preferred habitat and prey of each species of ladybird is very poorly documented and needs to be known in detail if we are to fully understand what determines prey specificty (cf. p. 98).

Thus although the quality of an aphid as food for a ladybird is important it is not the only factor. An aphid's availability and competition from conspecifics and other natural enemies are equally important. As the relative and absolute abundance of aphids and natural enemies vary from year to year the utilization of the various species of aphids in a habitat by a particular species of ladybird is also likely to change from year to year. That is, there is unlikely to be a rigid demarcation between essential and alternative prey, and whether an adult is searching for oviposition sites or for food to sustain itself further obscures such a demarcation. Following the thinking outlined above it is likely that each species of polyphagous ladybird oviposits in patches of a set of species of prey. The ranking of these nursery prey, as they will be referred to here, is not likely to be fixed as it is dependent on several factors some of which vary in time. When prey is generally abundant then ladybirds are likely to oviposit in colonies of a few preferred species of prey; however, when prey is uncommon they are likely to exploit more species of nursery prey (Fig. 5.9). The dietary range of ladybirds seeking to sustain themselves is likely to be much broader, often including plant material, and is referred to here as food prey or food.

SWITCHING

The majority of population models deal with monophagous predators. The response of polyphagous predators to a range of prey has largely been overlooked despite the fact that it is of considerable importance to predator foraging in the field. As indicated above even polyphagous ladybirds appear to prefer or do better when fed certain species of prey. However, if instead of showing a fixed preference for a particular prey species the proportion attacked changes from less than to greater than expected as the proportion of that prey available to the predator increases the predator is said to show switching. Such switching or fequency-dependent predation (Murdoch, 1969; Sherratt & Harvey, 1993) is likely to contribute markedly to the stability of prey populations if the predator population remains relatively constant (Hassell, 1978).

Conditions that favour switching in general predators are patchily

Fig. 5.9. The predicted relationship between number of prey species exploited and habitat quality.

distributed prey, and predators that are not restricted to one patch of prey and can detect prey at a distance (Cornell, 1976). Ladybird larvae have poor powers of dispersal, which restricts them to one patch of prey, and they can only determine the presence of prey over a short distance (p. 100). In addition mixed infestations of prey species are also relatively uncommon; therefore, it is unlikely that larvae have been selected to choose between prey species, even within a patch Their behaviour serves them well as it keeps them within a patch and close to aggregations of prey, which is adaptive as it enables them to maximize their net rate of food intake (p. 109). Therefore, one would not expect them to show switching. Switching, if it occurs in ladybirds, is likely to be shown by adults. Because of their greater powers of dispersal they can allocate a greater fraction of their searching time to those parts of the habitat more profitable for oviposition. That is, they may show an aggregative response not only to high-density patches of prey but also to those of a particular species.

The existence of switching can be demonstrated very easily and as a consequence several species of ladybirds have been so tested. No evidence was found for switching in adult *Stethorus punctorum* when presented with different proportions of two species of mites on discs cut out of apple leaves (Houck, 1986), or in adult *C. septempunctata* and *Harmonia axyridis* when presented with different ratios of aphids and mites on apple saplings (Lucas *et al.*, 1997*a*). Similarly for larvae of *C. septempunctata* foraging in mixtures of two aphid species, *Acyrthosiphon pisum* and *Aphis fabae* (Murdoch & Marks, 1973), and those of *Adalia bipunctata* foraging for three species of birch aphids (Hajek & Dahlsten, 1987). That is, these experimental studies confirm that larvae cannot be conditioned to a specific prey and tend to grasp and attempt to consume whatever the size or species of prey they encounter.

The only field study of switching in ladybirds is that done on the adults of *Chilocorus bipustulatus* feeding on scale insects on citrus in Israel (Mendel *et al.*, 1984). An analysis of the mid-gut contents of the adult ladybirds indicated that when the population density of young larvae of the Mediterranean black scale, *Saissetia oleae*, was high compared to that of the armoured scales, *Ceroplastes floridensis* and *Coccus hesperidum*, the predator switched to eating mainly black scale. Later in the year, as the black scale matured, the predator switched to eating mainly armoured scales. In this system the two types of prey are separated spatially with the early stages of the black scale mainly feeding on the leaves and the armoured scales mainly feeding on the branches of citrus trees. Similarly, adults of *Chilocorus nigritus* are thought to switch from exploiting *Aonidiella aurantii* on citrus to exploiting other diaspid scale insects on other species of plants (Samways, 1984, 1988). Thus, as predicted there is evidence that the adults of some ladybirds show switching. This is an aspect of ladybird ecology worthy of further study, especially in the field.

ADULT FORAGING BEHAVIOUR

Energy from prey is used by adults to fuel their searching behaviour and reproduction. The way in which ladybirds partition incoming energy between these two activities is unknown (Fig. 5.10). It is likely that the fat body is the main store of energy for searching and overwintering (Hodek & Honek, 1996). However, what priorities determine the partitioning of energy between fat body and gonads are unknown and need to be studied. There is some evidence that ladybirds emerging from hibernation may utilize some of their fat reserves to develop eggs (p. 29). Aphid populations tend to develop very rapidly in spring so to anticipate this could be adaptive. The results of the numerous studies on the use of non-aphid prey by aphidophagous ladybirds are confusing because one would expect them to develop and/or oviposit eggs only when they have located a patch of nursery prey. Some studies, like that of Richards & Evans (1998) indicate

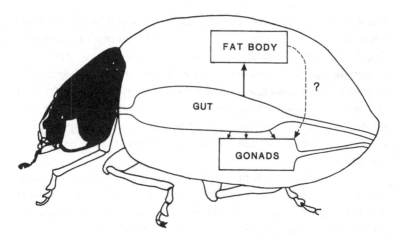

Fig. 5.10. Diagrammatic representation of the potential partitioning between fat body and gonads of the energy assimilated by adult ladybirds from their food.

that adults of *Coccinella septempunctata* and *C. transversoguttata* fed a diet of weevil larvae increase in weight but do not produce eggs. Others indicate that several species, notably *A. bipunctata, C. septempunctata, C. maculata* and *H. axyridis*, can be reared and will produce eggs on artificial diets that do not contain aphid material, and on the eggs of various lepidoptera (Kariluoto, 1980; Ferran *et al.*, 1997a; Phoofolo & Obrycki, 1997). In the absence of a choice it is not surprising that larvae eat and complete their development on such food. Indeed some like *C. maculata* fed the eggs of the moth *Ostrinia nubilalis* have a higher intrinsic rate of population increase than individuals fed pea aphids (Phoofolo & Obrycki, 1997). Thus, there is no doubting the high quality of such food. However, if a ladybird's offspring are to survive the prey has to be abundant for long enough to sustain their development.

Are these responses simply laboratory artifacts ? Field observations and laboratory studies (Dimetry & Mansour, 1976; Evans & Dixon, 1986) indicate that both *A. bipunctata* and *C. septempunctata* respond to aphid cues and so tend to oviposit close to aphid colonies, which is adaptive. However, ladybirds maturing and laying eggs on artificial diets (Kariluoto, 1980) would appear to be maladaptive. This paradox needs to be resolved, and its resolution is likely to improve our understanding of the way ladybirds partition their resources between fat body and gonads, and prey specificity.

Many adult ladybirds are known to have a wide dietary range but their larvae are only found developing on a few species of nursery prey

(Hodek, 1962, 1993; Mills, 1981*b*). In such cases developing eggs and ovi-positing in response only to the presence of large numbers of the nursery prey would appear to be adaptive. The difficulties researchers have experi-enced getting some ladybirds to mature eggs and oviposit on synthetic diets, which do not contain aphid material, tend to support this conten-tion. The fitness of females depends on their locating and ovipositing in high-quality patches of nursery prey, whereas that of males is dependent on their success in finding and fertilizing females. Therefore, the follow-ing account is restricted to an analysis of the searching behaviour of females. The factors determining the searching behaviour and fitness of males are discussed in Chapter 3 (p. 44).

Location of prey

Each species of ladybird tends to be associated with a particular habitat (Honek, 1985; Majerus, 1994). Thus in foraging for resources lady-birds are likely to respond to cues at three levels: those associated with habitat, patches of prey and individual aphids (Hassell & Southwood, 1978; Ferran & Dixon, 1993; Hattingh & Samways, 1995). In a seminal study Honek (1985) used nine variables that quantify environmental con-ditions to define niche overlap in seven species of ladybird (Table 5.1; p. 98). This is the first and only rigorous study of a subject that is funda-mental for understanding prey specificity, the structure of aphidopha-gous guilds and intraguild predation. Currently the habitats of most species are poorly known and where it is known there is very little infor-mation on how they locate them. That is, their habitats and location are the poorest understood aspects of adult foraging behaviour.

Adults appear to search mainly during the day. In the case of *C. sep-tempunctata brucki* this diurnal rhythm is determined by an endogenous circadian timer, which overrides hunger as a determinant of searching activity. That is, they readily eat at night but their searching activity is suppressed (Nakamuta, 1987). Visual responses are possibly important in the long-range location of habitat. In an arena *Chilocorus nigritus* is respon-sive to images of trees and specific leaf shapes (Hattingh & Samways, 1995) and *A. bipunctata*, a tree-dwelling species, orientates more markedly to the taller of two objects, whereas *Coccinella septempunctata*, which is mainly associated with herbaceous plants, does not differentiate between short and tall objects (Khalil *et al.*, 1985). Similarly, the adults of *H. axyridis* respond at a distance to geometric shapes more strongly than their larvae (Lambin *et al.*, 1996).

When searching in the field ladybirds do not move at random but

Table 5.1. Percentage niche overlap of seven species of ladybirds based on their abundance in a range of habitats in two years, 1982 and 1983, in the Czech Republic

			Percentage overlap			
Species	Coccinella septempunctata	Coccinella quinquepunctata	Propylea quatuordecimpunctata	Adonia variegata	Adalia bipunctata	Adalia decempunctata
Coccinella quinquepunctata	37					
Propylea quatuordecimpunctata	23	45				
Adonia variegata	50	16	4			
Adalia bipunctata	17	9	18	7		
Adalia decempunctata	6	11	18	4	15	
Calvia quatuordecimguttata	14	13	34	0	38	32

Note:
After Honek, 1985.

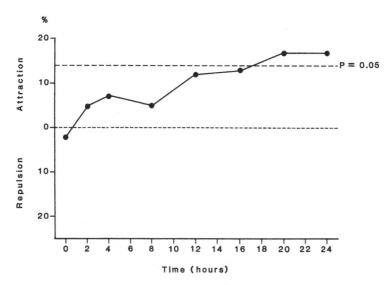

Fig. 5.11. The increasing positive response shown by adults of *Cryptolaemus montrouzieri* in an olfactometer to the odour of its prey, *Planococcus citri*, relative to the period the adults have been kept without food. Upper dashed line indicates a probability of occurring by chance of $P = 0.05$. (After Sengonca *et al.*, 1995.)

appear to be attracted to sites where prey are or have recently been present. For example, *Rodolia cardinalis* quickly discover patches of the cottony-cushion scale, *Icerya purchasi*, usually within a fortnight of their being established artificially. The proximity of these patches to sources of ladybirds did not affect the speed with which they were found (Prasad, 1990). In the laboratory prey odour is attractive to predators (Colburn & Asquith, 1970; Obata, 1986). Olfactometer experiments indicate that *Chilocorus nigritus*, *Coccinella septempunctata* and *Cryptolaemus montrouzieri* are all attracted by the odour of their prey. In the case of *C. montrouzieri* this response is most marked when adults are hungry (Fig. 5.11) In addition to being attracted by the odour of their prey adults also indulge in intensive search (area-restricted search), i.e. they move more slowly and turn more frequently in the presence than in the absence of the odour of prey (Heidari & Copland, 1992; Sengonca & Kotikal, 1994; Sengonca & Liu, 1994; Hattingh & Samways, 1995; Ponsonby & Copland, 1995; Sengonca *et al.*, 1995).

On locating a patch of prey it is likely that adults respond to cues associated with their prey and as a consequence search the area more thoroughly. For example, the wax of mealy bugs is an arrestant stimulus

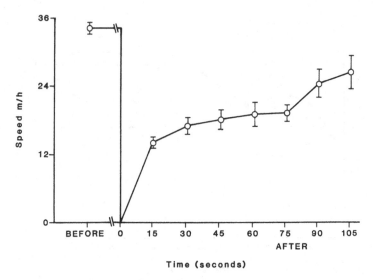

Fig. 5.12. The speed of movement of adult *Hippodamia convergens* before and at intervals after feeding on an aphid. (After Rowlands & Chapin, 1978.)

for *Diomus* and *Exochomus* (van den Meiracker *et al.*, 1990), and *C. montrouzieri* (Merlin *et al.*, 1996a). In the absence of the wax filaments of its prey *C. montrouzieri* will develop eggs but not oviposit (Merlin *et al.*, 1996a). In addition they can detect and orientate to individual prey prior to actual physical contact. In simple arenas *Coccinella septempunctata* can do so up to a distance of 1 cm and *Anatis ocellata*, 1.3–1.9 cm (Allen *et al.*, 1970; Stubbs, 1980; Nakamuta, 1984). Evidence that visual cues may play a role comes from laboratory studies in which *C. septempunctata* consumed more of the one of two colour morphs (red and green) of the pea aphid that contrasted most with the background colour and that *H. axyridis* consumed more of the red morph irrespective of background colour (Harmon *et al.*, 1998). Although it is not known to what extent olfactory and visual cues are used in this orientation it is clear the response only occurs over a relatively short distance. This possibly accounts for why so many authors have claimed that ladybirds only respond to the presence of prey after physical contact (e.g. Fleschner, 1950; Banks, 1954, 1957; Dixon, 1959; Kehat, 1968; Storch, 1976).

Virtually all predators search more thoroughly for food in some areas than they do in others. Adult ladybirds conform to this in that having fed on a prey item they then often move more slowly and turn more frequently (Figs. 5.12, 5.13; Rowlands & Chapin, 1978; Podoler & Henen, 1986; Kareiva & Odell, 1987). This non-random foraging, or area-restricted search, results in their staying in the vicinity for longer, which

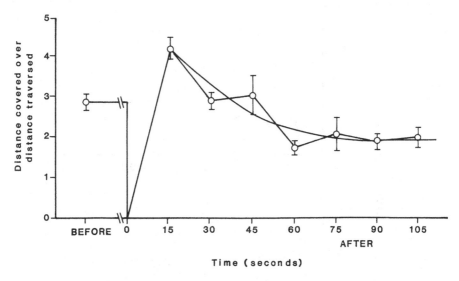

Fig. 5.13 The thoroughness of search of adult *Hippodamia convergens* before and at intervals after feeding on an aphid. (After Rowlands & Chapin, 1978.)

if the prey is clumped should result in an increase in the rate of prey capture. The switch from extensive to area-restricted search is elicited by contact with rather than consumption of prey, and the time spent in area-restricted search is positively correlated with the time spent feeding on the most recently consumed prey (Fig. 5.14; Nakamuta, 1985*a*, *b*). In addition, there is evidence that area-restricted search only occurs after eating certain prey. For example, adults of the two-spot ladybird *A. bipunctata* switch from extensive to intensive search after capturing an hop aphid, *Phorodon humuli*, but not after feeding on low-quality prey like *Aphis fabae*, *A. craccivora* or *A. spiraephaga* (Kalushkov, 1998, 1999). That is, the switch to area-restricted search depends not only on locating and capturing prey but also on the quality of the prey, and also possibly on prey abundance and whether a ladybird is seeking nursery or food prey.

When caught by a predator aphids often release an alarm pheromone, which causes aphids close by to cease feeding and leave the area, or in the case of the few species of aphids that have soldiers the alarm pheromone attracts soldiers, which then attack the predator (Arakaki, 1989). However, there is no evidence that adults of ladybirds like *C. septempunctata*, which feed on aphids that do not have soldiers, utilize alarm pheromone to locate aphid prey (Nakamuta, 1991).

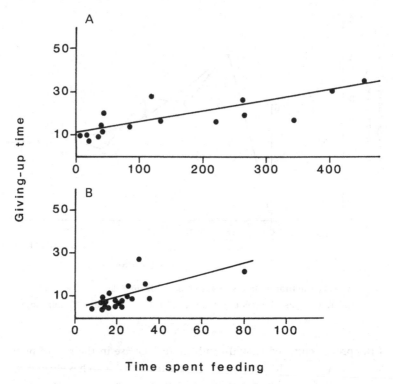

Fig. 5.14. Time spent in area-restricted search (giving-up time) by adults of *Coccinella septempunctata* relative to the time previously spent consuming (A) artificial prey and (B) the aphid *Myzus persicae*. (After Nakamuta, 1985*b*.)

Patch quality

The quality of patches of nursery prey varies not only in terms of whether the prey is suitable but also in terms of their dynamics in time. The importance of this within-patch dynamics differs in the two major groups of predatory ladybirds.

Even in the absence of natural enemies aphid colonies initially increase very rapidly in abundance and then may equally rapidly decline in abundance when the aphid switches to producing winged individuals and disperses to establish new colonies elsewhere. The time-scale over which this occurs is very similar to the time it takes an aphidophagous ladybird to complete its immature development (Fig. 5.15). In addition, if a patch is exploited by a large number of ladybirds and other aphidophaga then the dynamics of the prey is likely to change; the peak abundance could be lower and the decline in abundance could occur earlier (Fig. 5.15(C)). Another feature of the aphid–ladybird interaction, which may

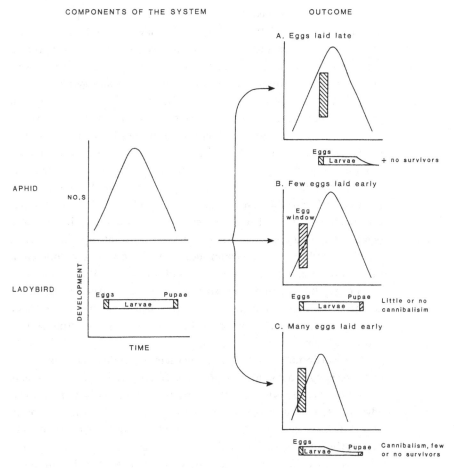

Fig. 5.15. Graphical presentation of the components of the ladybird–aphid interaction: temporal changes in the abundance of aphids and relative developmental time of the ladybird, and the outcome if (A) the eggs are laid late, (B) a few eggs are laid early, or (C) many eggs are laid early.

also introduce further uncertainty, is that some aphids appear to be able to respond to the presence of ladybird larval tracks and switch to producing winged forms, which leave the patch before the ladybird larvae reach a large size and pose a serious predation risk to the aphids (Dixon & Agarwala, 1999). This could have disastrous consequences as the aphids become scarce before the ladybirds can complete their development. Therefore, one would expect them to lay a few eggs early in the development of a patch (Fig. 5.15(B)). However, if they lay eggs too early then the population density of the aphid is likely to be too low for their first instar larvae to be able to catch sufficient aphids to survive (Dixon, 1959).

It is likely, but needs to be established, that an aphidophagous lady-bird matures most of the eggs it is going to lay in a patch by feeding on aphids in that patch and only stays and lays eggs if its rate of capturing aphids exceeds a certain critical value. Adults, relative to their large larvae, are very inefficient at capturing aphids (Dixon, 1959). This could be viewed as adaptive as it makes it less likely that they will oviposit in patches where the abundance of prey is too low to sustain the develop-ment of their small larvae. The inefficiency of the adults is mainly due to their domed shape, in profile, and aposematic colouring. If the adults had a more flattened shape and were more cryptically coloured then it is likely they would be more efficient at capturing prey and would start ovi-positing earlier in the development of an aphid patch. In this situation the survival of the first instar larvae would be enhanced if they were bigger. Assuming adult size and the larval growth rate does not change then this would appear to have an added advantage as the developmental time would be shorter. However, the disadvantage is that they would lay fewer eggs. Thus the colour and shape of an adult could also be viewed as adaptive in terms of foraging (p. 33). This aspect of the ladybird–prey interaction is poorly understood and likely to prove an interesting optimi-zation problem.

Most coccidophagous species lay their eggs individually under the ovisac or body of a coccid. The larva that emerges from the egg first eats the contents of the ovisac and then the adult coccid. The larva reaches an advanced stage of development on one prey item. That is, the availability of prey is more certain than it is for aphidophagous ladybirds. In addi-tion, as the egg is placed immediately under an immobile coccid the larva does not have to hunt for and subdue its prey. However, even here it would appear to be advantageous to oviposit under a coccid that is at a particu-lar stage in its development. Late on in its development most of the coccid's offspring will have dispersed; if too early, then the larva will have to eat the adult coccid before it has produced all its offspring. That is, as in the aphidophagous species, it is likely that coccidophagous species are also faced with an optimization problem when selecting where to ovi-posit. Little work has been done on the oviposition behaviour of cocci-dophagous species. There is some evidence that *Chilocorus nigritus* might avoid heavy infestations of red scale because they are unlikely to last long enough to sustain the development of their larvae (Ericksen *et al.*, 1991), and stronger evidence that chemical cues left by larvae of *Cryptolaemus montrouzieri* deter the adults from ovipositing (Merlin *et al.*, 1996b). In view of the poorer level of knowledge on coccidophagous ladybirds the follow-ing account is restricted to aphidophagous ladybirds.

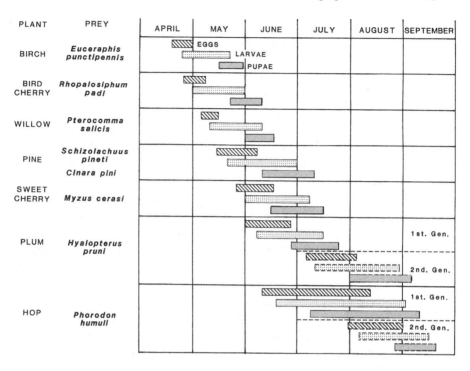

Fig. 5.16. The distribution in time of the eggs, larvae and pupae of *Adalia bipunctata* in colonies of aphids on different species of trees in a woodland habitat. (After Pruszynski & Lipa, 1970.)

Assessment of patch quality

As indicated above the patches of prey that serve as nurseries for the larvae of aphidophagous ladybirds are ephemeral in space and time. As a consequence adults of generalist ladybirds, like *Adalia bipunctata*, exploit a sequence of species of aphids during the course of a season, and only produce two generations in long-lasting patches of aphid prey, such as those of the hop and plum aphids (Fig. 5.16). An exception to this are the breeding grounds of the ladybirds that have a close association with ants (Wheeler, 1911; Berti *et al.*, 1983). For example, the rare seven-spot, *Coccinella magnifica*, is always found in association with wood ants (Majerus, 1989). Ants like *Formica polyctena* palpate the eggs and larvae of this ladybird but rarely attack them, whereas they immediately attack and destroy the eggs and larvae of the similar-sized *C. septempunctata*. Interestingly, adults of *C. magnifica* are attacked, but because of their shape, the ant finds them more difficult to grip with their mandibles than the adults of *C. septempunctata* (Godeau & Hemptinne, 2000). The adults of *C. magnifica* remain in the vicinity of an ants' nest all year round and breed when the aphids attended by the ant become abundant. The eggs are usually laid some distance from an aphid colony, which may also account for why this species lays relatively large eggs (cf. p. 69). The hatchling larvae have to locate an aphid colony and the bigger it is the greater the probability of it doing so. That is, the locations of breeding grounds are more fixed in space and long-lasting/predictable. Therefore, there are advantages in the ladybird staying with the 'resource' all year round. This raises interesting questions about how this ladybird regulates its numbers relative to the carrying capacity of each nursery area, which would repay further study.

Having located a patch of prey the problem for most ladybirds is whether the patch is suitable for their offspring. In view of the ephemeral nature of the resource it is possible that ladybirds use the age structure of the aphid colony, as has been reported for syrphids (Kan & Sasakawa, 1986; Kan, 1988a, b), or the phenological age of the plant, both of which are often correlated with the age of an aphid colony – an aphid colony with a predominantly young age structure on a plant that is at an early stage of development is likely to last longer than an aphid colony that is predominantly made up of old aphids on a maturing plant. This hypothesis has been tested experimentally in the laboratory using *A. bipunctata*. Neither the age of the host plant or the age structure of the aphid colony appeared to affect oviposition behaviour (Hemptinne *et al.*, 2000c).

Theory indicates that these ladybirds should lay eggs early in the development of a patch of aphids (Fig. 5.13(B); Kindlmann & Dixon, 1993). Experimental and field studies indicate that there is a density below which ladybirds are unlikely to lay eggs (Dixon, 1959; Wratten, 1973; Honek, 1978; Mills, 1979). In addition, in the field two-spot ladybirds tend to lay their eggs well before aphid populations peak in abundance (Fig. 5.17). That is, there is a window in the development of a patch of aphids when ladybirds are most likely to lay their eggs. The opening of the window is possibly determined by the minimum density of aphids required for the survival of the first instar larvae. The closing of the window appears to be initiated by the adults' response to the tracks left by conspecific larvae. In the presence of such tracks gravid females of *A. bipunctata* and *C. septempunctata* become very active and if prevented from leaving the area refrain from laying eggs for a few hours. The deterrent effect is density dependent and mediated via a pheromone present in the larval tracks. That is, larvae of ladybirds produce a species specific

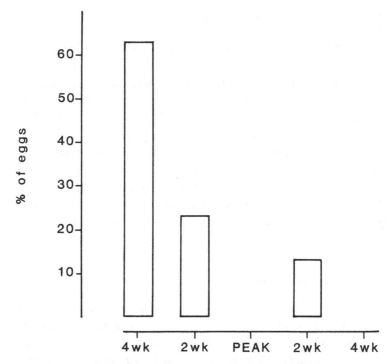

Fig. 5.17. Distribution in time, relative to peak aphid abundance, of the laying of eggs by *Adalia bipunctata* on lime trees. Development of aphid populations expressed in weeks before and after the recorded peak in aphid abundance in each year. (After Hemptinne *et al.*, 1992.)

oviposition-deterring pheromone. In the case of the two-spot ladybird the oviposition-deterring pheromone consists of a cocktail of alkanes in which methyl-branched (C23 to C33) and straight-chain alkanes (C20 to C31) predominate. The fact that the oviposition-deterring pheromone consists of a blend of alkanes is interesting for two reasons. Firstly, these molecules spread easily on the hydrophobic cuticle of plants and so leave a large signal. Secondly, they are stable, with the inhibiting effect on oviposition lasting for at least 10 days (Hemptinne *et al.*, 1992; Doumbia *et al.*, 1998; Hemptinne *et al.*, 2000e). Růžička (1997a) has similarly shown that the deterrent effect of the oviposition-deterring pheromone of *Chrysopa oculata* persists for 3 to 4 weeks and is not degraded even by keeping it at a temperature of 140 °C for 60 minutes.

Although it is unrealistic to equate laboratory and field conditions, the results of these studies complement what is known of ladybird ecology. The eggs and young larvae are the two most vulnerable stages of these insects. Of all the potential enemies, ladybird larvae are probably the most important threat. Egg cannibalism is strongly density dependent (Mills, 1982a) and the later in the development of an aphid colony oviposition occurs the more intense is non-sibling cannibalism (Osawa, 1989). Not ovipositing in patches of prey where conspecific larvae are already present reduces both the hazard of cannibalism and competition for food. In addition the presence of ladybird larvae could also indicate that the aphid colony is in a late stage of development. That is, by responding to aphid abundance and cues indicating the presence of conspecific larvae adult ladybirds are more likely to oviposit in aphid colonies that will support the development of their offspring. Current studies indicate that an adult's experience of average patch quality and distance between patches (transit time) can also influence whether its stays and oviposits or leaves a patch.

EGG DISTRIBUTION

One striking difference between aphidophagous and coccidophagous ladybirds is that the former lay clusters of eggs whereas the latter predominantly lay eggs singly, which are often placed beneath prey and therefore not easily accessible to other predators. The laying of eggs in clusters by aphidophagous ladybirds has been viewed as a defence against heterospecific predators, because they are more deterred from attacking clusters than single eggs. This is thought to be due to the presence of alkaloids in the eggs, the deterent effect of which is thought to be dose dependent (Agarwala & Dixon, 1993). That the primary reason for the evolution of egg clustering in aphidophagous ladybirds is defence, however, is weakened by the observation that alkaloids are also present in the eggs of coccidophagous species (Pasteels et al., 1973).

Another possibility is that egg clustering facilitates 'social feeding'. When larvae disperse from an egg cluster they are very vulnerable and their survival is dependent on quickly obtaining their first meal (Dixon, 1959). Observations on larvae during this period indicate that they orientate towards larvae that have already captured an aphid, which often results in several larvae feeding on the same aphid. The larvae do this by orientating to the 'alarm pheromone' released by the captured aphid and searching the area thoroughly (Hemptinne et al., 2000d). That is, by laying clusters of eggs these ladybirds may be increasing the probability of a larva quickly obtaining its first meal and so surviving to the next instar.

LARVAL FORAGING BEHAVIOUR

Theory indicates that the fittest adults are those that oviposit in patches of nursery prey at an early stage in their development (Kindlmann & Dixon, 1993). That is, larvae are most likely to hatch close to an abundance of prey. Thus selection will favour those larvae that stay in the patch close to clusters of prey. The generally non-cryptic colouring of the larvae of many aphidophagous ladybirds is surprising as it is likely to reduce their efficiency in capturing prey. As the larvae also contain toxic alkaloids it is likely this colouration is aposematic and serves to warn off visually hunting predators like birds. High densities of aphids are also attractive to many species of birds and under these circumstances it could be advantageous for the larvae to advertise their toxicity even if it makes them less effective as predators. The general similarity of larval colouring between species might indicate they may also benefit from Müllerian mimicry (p. 32).

Location of prey

As with adults the larvae are most active during the day with a peak in activity occurring early in the afternoon (Fig. 5.18). They also tend to be positively phototactic and negatively geotactic, which results in them generally climbing up through vegetation to the growing points where their prey are most frequently found. In addition, they tend to follow the edges of leaves and walk along prominent leaf veins, which similarly often brings them into closer proximity with their prey (Dixon, 1959).

Larvae that have just fed on an aphid tend to walk more slowly and to change their direction of movement more frequently than they did just prior to encountering an aphid (Dixon, 1959; Kawai, 1976; Hunter, 1978). Similar results have been obtained for two species of coccidophagous ladybirds (Podoler & Henen, 1986). That is, larvae like adults (Figs. 5.12, 5.13; p. 100) more thoroughly search the area immediately around where they capture prey, i.e. they also show area-restricted or intensive search. If unsuccessful in locating further prey then the thoroughness of search decreases (Fig. 5.19) and speed of movement increases (Fig. 5.20) with time. As a consequence of a larva switching from area-restricted search to extensive search it leaves the area, and within a short time at a rate in terms of linear distance per unit time equivalent to that observed before the encounter (Fig. 5.21). That area-restricted search is adaptive has been elegantly shown by Murakami & Tsubaki (1984). By arranging the same number of aphids uniformly or in four clumps at two levels of

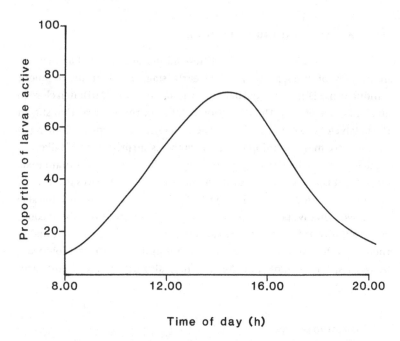

Fig. 5.18. The diurnal cycle of activity shown by hungry fourth instar larvae of *Coccinella septempunctata*. (After Carter, 1982.)

aggregation they showed that the number of aphids eaten per unit time by fourth instar larvae of *C. septempunctata* is greatest when prey distribution is highly aggregated. In addition, Kawai (1976) has shown that this behaviour also results in larvae of *Harmonia axyridis* remaining longer on heavily infested than on lightly infested plant shoots (Fig. 5.22). That is, larvae are well adapted to exploiting clumped prey.

Encounters with prey that do not result in a capture also initiate a change in searching behaviour (Fig. 5.20). Although the general form of the response to an unsuccessful encounter is similar to that observed after a successful encounter there is a difference in the search speed and thoroughness immediately after the encounter, probably attributable to larval motivation. After losing contact with prey larvae appear more intent initially on pursuit rather than trying to find another prey item. Contact with cues associated with aphids, such as honeydew, can also induce larvae to change their searching behaviour. In the presence of honeydew larvae spend longer searching an area, resulting in a greater exploitation of prey, than in the absence of honeydew. Larvae, however, do not respond to differences in the amount of honeydew. As honeydew is rarely confined to the immediate vicinity of prey and in many cases falls

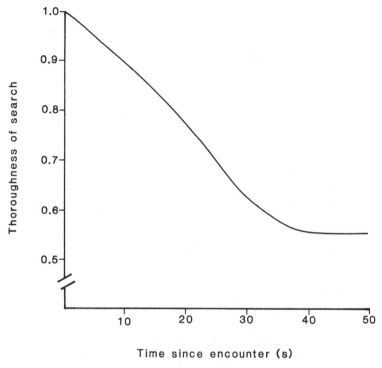

Fig. 5.19. Change in the thoroughness of search in time shown by fourth instar larvae of *Coccinella septempunctata* after eating an aphid. (After Carter & Dixon, 1982.)

or is thrown onto parts of plants where prey are absent, such as the upper surfaces of leaves, this arrestant response to honeydew only results in the larvae aggregating in areas where prey are present (Carter & Dixon, 1984).

It is claimed that larvae of *C. septempunctata* can respond to their own tracks and so avoid searching areas they have previously searched. This is based on a decline in the time spent searching the same plant in each subsequent visit, which is reversed after the plant is carefully cleaned with acetone (Fig. 5.23; Marks, 1977). However, the length of time spent searching after washing with acetone is considerably less than on the first visit and is not followed by a declining trend in search times. Attempts to repeat this study by observing the behaviour of *C. septempunc-tata* larvae crossing an area they have previously traversed several times revealed no indication that they respond to their own or other individual's tracks (unpublished results). That ladybirds set to search a vertical object spend less time doing so on each subsequent visit has been attributed to increasing hunger (Dixon, 1959). As it has not been possible to

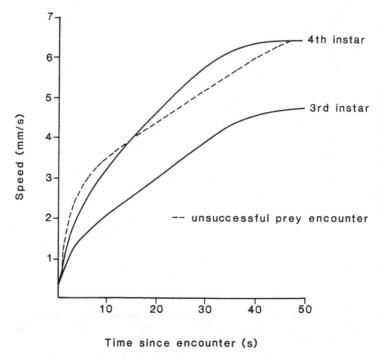

Fig. 5.20. Change in the speed of movement in time shown by third and fourth instar *Coccinella septempunctata* after eating an aphid (——) and after encountering but failing to catch an aphid (- - - -). (After Carter & Dixon, 1982.)

repeat Marks's (1977) observations, hunger plus a temporary heightened response to a novel substrate is possibly the best interpretation of this data set. That the supposed signal is specific to individual larvae also posses problems when considering the mechanism and possible adaptive significance. This is an aspect of foraging behaviour that would repay further study.

Within patches of prey there are likely to be spatial differences in prey abundance, e.g. if one regards a tree as a patch then there are very big differences in the numbers of aphids present on each of the leaves. That is, within a patch prey distribution is likely to be clumped. In response to this clumping of prey larvae of invertebrate predators forage longer and more intensively where prey is abundant and leave areas where prey is scarce (cf. Fig. 5.22; Hassell, 1978). In developing optimal foraging models for predators like ladybird larvae it is important to appreciate the morphological and physiological constraints that are likely to affect their searching behaviour. Prey detection by larvae appears only to occur on

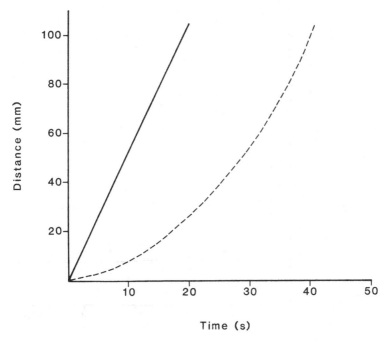

Fig. 5.21. Change in linear displacement of larvae of *Coccinella septempunctata* from feeding sites (- - -). The solid line represents linear displacement if larvae do not show area-restricted search. (After Carter & Dixon, 1982.)

physical contact or over short distances when prey are crushed (Stubbs, 1980) and there is no evidence to suggest that prey density can be determined by odour or honeydew concentration (p. 110). If larvae are unable to estimate variations in prey abundance are they capable of foraging efficiently in patches where there are spatial differences in prey abundance?

The foraging process of coccinellid larvae has three components: prey is subdued and eaten (handling time), followed by a short period of intensive searching, followed by a period of extensive searching by which a larva moves between clumps of prey within a patch. Hunger has been shown to affect many components of the foraging process (Hassell *et al.*, 1976) and has been suggested as probably the best indicator of the average feeding rate available to invertebrate predators (Charnov, 1976*a, b*).

After feeding on an aphid hungry larvae of *C. septempunctata* search more slowly and change direction more frequently, which results in a more intensive search around the feeding site. This greatly increases their chances of encountering an adjacent aphid. When in the extensive search mode hunger in this species does not appear to influence the speed of movement or time spent searching a particular plant structure.

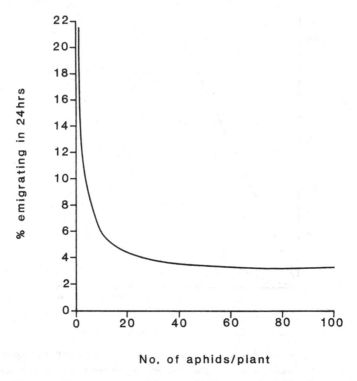

Fig. 5.22. The percentage of larvae of *Harmonia axyridis* leaving cereal seedlings within 24 hours relative to the number of aphids (*Rhopalosiphum padi*) infesting these plants. (After Kawai, 1976.)

When set to search for prey in a simple tree like system consisting of four sub patches (Fig. 5.24) the hunger level of a larva and the density of prey markedly affect the outcome. When a prey item is encountered, the duration of intensive search, by affecting the probability of encountering further prey, influences the time larvae spend in a sub patch (Fig. 5.25), and ultimately the degree to which each sub patch is depleted (Fig. 5.26). In sub patches where there were few prey, however, initial encounters rarely lead to subsequent encounters so that time in the sub patch is unaffected by hunger. In addition hungry larvae have a longer handling time than well-fed larvae (Fig. 5.27).

Area-restricted searching behaviour is a simple mechanism that enables a predator to exploit patchily distributed prey. Its duration is probably related to the predator's search rate and the average sub patch size of its prey (Hassell & Southwood, 1978), and has usually been regarded as of constant duration. This results in all but the lowest density sub patches being left at the same final density. This is clearly inefficient

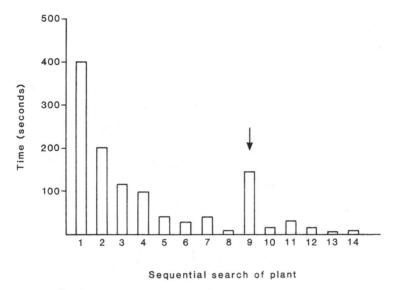

Fig. 5.23. Search times resulting from sequential introductions of *Coccinella septempunctata* larvae to an uninfested bean plant. The arrow indicates the stage in the sequence when the plant was cleaned with acetone. (After Marks, 1977.)

where prey availability within the patch changes in time (cf. p. 103). In such circumstances one would expect the time spent in each sub patch to change as their average quality changes. The longer intensive searching period of hungry larvae increases their probability of encountering a subsequent prey item, leading to longer sub patch times and greater exploitation, i.e. sub patches are left at lower final densities.

This raises an interesting question: how do larvae measure sub patch quality? Do they measure the profitability of a patch over a period of time, as an average of the rate of gain of a particular benefit they are trying to maximize? Or is their measure an instantaneous estimate of the average rate of gain? Measures of patch quality fall into three main categories, each a function of the benefit the predator is assumed to be maximizing: average rate of encounter with prey, average rate of food harvested (mg/h), or average net rate of energy intake (Charnov, 1976*a*, *b*; Cook & Hubbard, 1977; Cook & Cockrell, 1978; Krebs, 1978; Commins & Hassell, 1979; Townsend & Hughes, 1981). Such estimates of profitability over time led to the postulation of a memory window and the speculation on its length, e.g. how many sub patches are included in the estimate (Krebs, 1978). Or, bearing in mind the constraints larvae are subject to, is their behaviour simply determined by hunger.

Top view

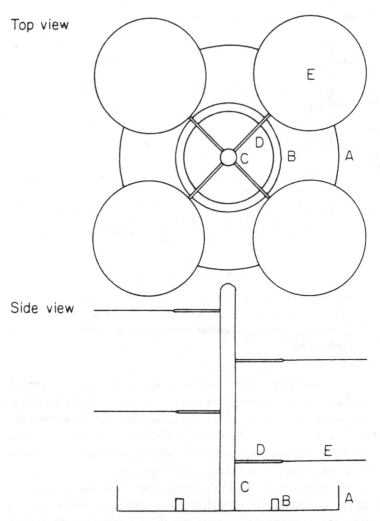

Side view

Fig. 5.24. Diagram of the apparatus used for studying the searching behaviour of larvae within sub patches of prey. (A) Petri dish; (B) Fluon-coated ring; (C) 9 cm long, 0.5 cm diameter wooden stem; (D) 3.5 cm long wooden petiole; (E) 5.5 cm diameter cardboard sub patch. (After Carter & Dixon, 1982.)

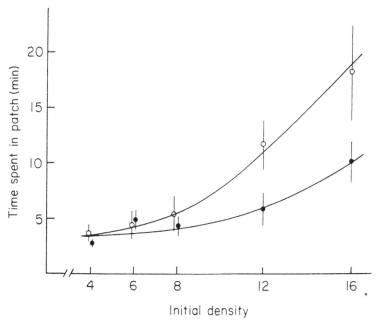

Fig. 5.25. Time spent in sub patches of different aphid density by larvae previously deprived of food for 5 hours (●) and 25 hours (○) respectively. (After Carter & Dixon, 1982.)

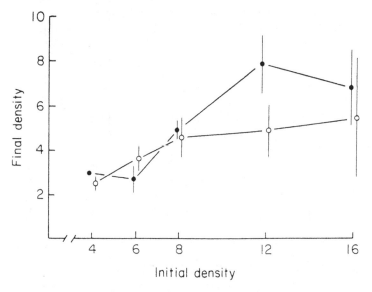

Fig. 5.26. The number of aphids remaining in sub patches of different initial aphid densities after visits by larvae previously deprived of food for 5 hours (●) and 25 hours (○) respectively. (After Carter & Dixon, 1982.)

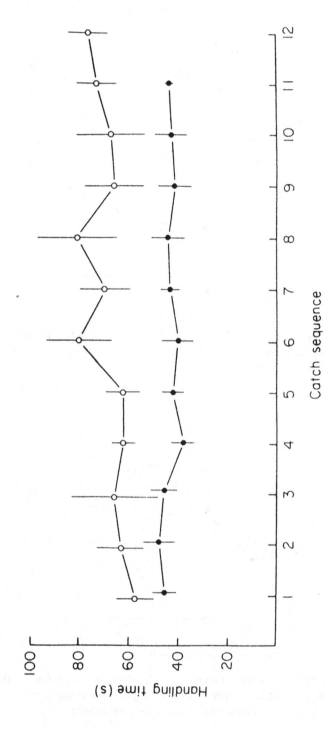

Fig. 5.27. The average time taken to handle each aphid in a sequence of prey captured by larvae previously deprived of food for 5 hours (●) and 25 hours (○) respectively. (After Carter & Dixon, 1982.)

Hunger or encounter rate

To simulate different conditions of sub patch quality and establish known encounter rates and hunger levels Carter & Dixon (1982) subjected *C. septempunctata* larvae to one of the following feeding regimes:

Group A – fed one first instar aphid (0.12 mg) followed by another every 7 min for 147 min, i.e. 22 aphids totalling 2.64 mg or 63% of the food required for satiation;

Group B – fed one fourth instar aphid (1.8 mg) followed by a first instar aphid every 21 min for 147 min, i.e. eight aphids totalling 2.64 mg or 63% of the food required for satiation;

Group C – fed one first instar aphid followed by another every 21 min for 147 min, i.e. eight aphids totalling 0.96 mg or 23% of the food required for satiation.

After the last meal their searching behaviour was monitored. Larvae in groups B and C experienced the same encounter rate, which was lower than that experienced by larvae in group A. If the area-restricted searching behaviour is determined by encounter rate then larvae of groups B and C should have similar but longer periods of intensive search than larvae of group A, i.e. $B = C > A$. The level of hunger, however, was the same for larvae of groups A and B , and lower than that of larvae in group C. If hunger determines the area-restricted searching behaviour then the duration of intensive search should be similar in larvae from groups A and B and shorter than that of the larvae in group C, i.e. $A = B < C$, which is what is observed (Fig. 5.28). Furthermore, encounter rate is only appropriate if each encounter imparts the same value to the estimate. It is not surprising, therefore, that the area-restricted searching behaviour of coccinellid larvae , which encounter a wide range of prey sizes, is not determined by average encounter rate. Similarly, area-restricted searching behaviour is not determined by the average rate of harvesting food as larvae of groups B and C have the same harvesting rate but differ in the duration of their area-restricted searching. This is because this rate and hunger level are not directly related. Hunger level is dependent not only on the amount of food eaten but also on the time that has elapsed since feeding. A standard prey item, however, has the same value in terms of weight eaten per unit time anywhere within the memory window used by a larva to measure harvest rate. Consequently, a specific hunger level could result from different rates of food intake (Holling, 1966). The third measure of patch quality, average net rate of energy intake, has not been measured for ladybird larvae. Although a larva trying to maximize its net

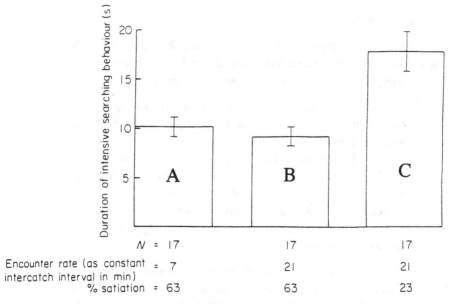

Fig. 5.28. The duration of intensive searching behaviour of larvae that experience different encounter rates and levels of hunger. For details of treatments A, B and C see text (p. 119). (After Carter & Dixon, 1982.)

rate of energy gain should use this measure if it is to forage optimally,
there is no evidence to suggest that they have the capability. It possibly
requires an implausible degree of omniscience on the part of a ladybird
and certainly a problem of detailed costing on the part of an investigator.
That is, hunger level appears to be the main determinant of searching
behaviour.

In some predators handling time decreases with increase in prey
density (Hassell *et al.*, 1976) or decreasing hunger (Johnson *et al.*, 1975;
Sabelis, 1981). This has led to the development of an optimal foraging
model in which handling time is determined by average encounter rate,
because extraction of food from a prey item becomes progressively more
difficult. At high encounter rates prey should be abandoned when it
becomes more profitable to capture another (Cook & Cockrell, 1978). Cook
& Cockrell also proposed a model based on hunger, the gut limitation
hypothesis, in which handling time is dependent on the intercatch interval because as this interval increases more space becomes available in the
crop as food passes into the mid-gut. That the capacity of the gut/crop
imposes a limit on the amount of food extracted from each prey item is
well illustrated by Mills's (1982) study of the larvae of *Adalia bipunctata*, in
which he varied the prey-to-predator weight ratio by feeding larvae of

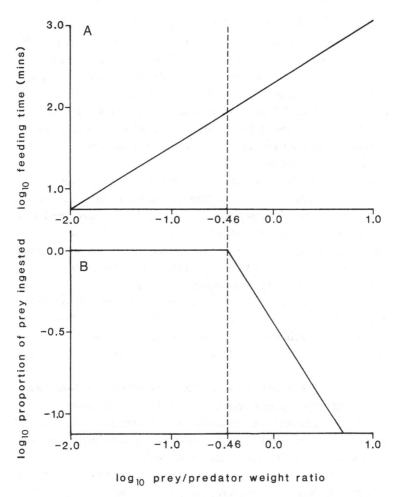

Fig. 5.29. The relationship between (A) feeding time and (B) proportion of prey ingested, and the prey/predator weight ratio (−0.46 is equivalent to a prey/predator weight ratio of 0.35). (After Mills, 1982b.)

each of the four instars aphids of various sizes. The time they spent feeding on each prey item is a simple linear function of the size of the prey relative to that of the predator; however, the proportion of each prey ingested is not. Relatively small prey are completely ingested but the proportion ingested of prey that are 0.35 or a greater proportion of the predator's weight falls from unity (Fig. 5.29). The break point in the relationship coincides exactly with the gut capacity (G) determined by feeding hungry larvae of different weights (W) to satiation with excess prey (G = 0.35W). However, the hungry larvae of C. septempunctata spent longer handling

prey than better-fed larvae (Fig. 5.25), even though their average encounter rates with prey were similar and the small prey were each totally
consumed. Therefore, an optimal foraging model based on average
encounter rates and the gut limitation model do not appear to apply to
ladybirds. In addition, the gut limitation hypothesis is implausible
because it is doubtful if a predator would forage when almost satiated
(Carter, 1982). However, if larvae assess prey availability by their level of
hunger instead of encounter rate then the predictions of the optimal foraging model are supported.

The longer handling times of hungry larvae possibly result from
the larvae spending longer searching the substrate with their mouthparts to ensure that each prey item has been totally consumed. This is consistent with the increased extraction of food from large prey items in
poor-quality habitats predicted by the optimal foraging model. However,
it is difficult to understand why handling time does not decrease through
a catch sequence as hunger decreases (Fig. 5.25), unless larvae measure
hunger by the contents of their mid-gut and not their crop, resulting in a
delay in their response to prey availability. This is again consistent with
optimal foraging theory because a predator should forage in response to
the average quality of the habitat and not to conditions currently experienced.

Although hunger level had no effect on extensive searching behaviour of C. septempunctata (Carter & Dixon, 1982) increasing hunger has been
shown to reduce turning rate and/or increase the searching speed of
extensive searching in larvae of A. decempunctata, leading to dispersal
from areas of poor prey availability (Dixon, 1959).

In conclusion, although many of the components of the foraging
behaviour of ladybird larvae have been shown to be hunger-dependent
this is an area largely ignored in the development of optimal foraging
theory. However, it is likely that it is by means of these hunger-induced
changes in foraging behaviour that ladybird larvae forage in a manner
similar to that predicted by optimal foraging theory and so maximize
their rate of energy intake.

Plant structure

Many invertebrate predators of aphids search along the prominent
contours of plants. The structure of a plant is important in determining
whether or not this behaviour leads a predator to the preferred feeding
site of its prey (Fleschner, 1950; Banks, 1957; Dixon, 1959; Boldyrev &
Wilde, 1969; Wratten, 1973). Similarly, the rate at which a predator

encounters prey can also be determined by characteristics of a plant such as hairiness and waxiness. Banks (1957) found that *Propylea quatuordecimpunctata* moved more slowly on hairy potato leaves than on glabrous bean leaves. Similarly, *Delphastus pusillus* consumes more whitefly per unit time and lays more eggs on equally infested cultivars of poinsettia with few trichomes on the undersides of their leaves (Heinz & Parrella, 1994). However, not all species of ladybird are similarly affected by the hairiness of a particular plant. For example, the larvae of *Coleomegilla maculata* are irritated by the glandular hairs of cucumber and frequently fall from the leaves whereas those of *A. bipunctata* and *Cycloneda sanguinea* do not (Gurney & Hussey, 1970). Similarly the wax bloom on the surface of the leaves and stems of pea plants (*Pisum sativum*) impairs the adhesion and mobility of adults of *Hippodamia convergens* . Both in the field and laboratory this can result in the ladybird capturing and eating more pea aphids (*Acyrthosiphon pisum*) when foraging on lines with a reduced wax covering compared to normal lines (Eigenbrode *et al.*, 1998).

The effect of plant structure on searching efficiency is well illustrated by the study of Carter *et al.* (1984). They calculated the functional response curves for third instar larvae of *Coccinella septempunctata* foraging for the same densities of pea aphid, *Acyrthosiphon pisum,* on pea and bean plants (Fig. 5.30). Transformation of these functional responses enables the attack coefficient (p. 82) to be calculated by linear regression of the logarithm of the proportion of prey surviving and the number of prey eaten. The intercept provides an estimate of the attack coefficient (Fig. 5.31). The lower attack coefficient on pea (0.45) than on bean (1.24) plants shows that peas are searched less efficiently than beans. This is not due to a difference in the coincidence of prey distribution and predator searching efficiency on the two plants, but to the larvae falling off the smooth leaves of pea plants more frequently than off the leaves of bean. Other studies have similarly shown that aphids are at greater risk of predation on certain plants and that this is primarily due to the plants' traits affecting the predators' residence times. In particular, the tendency for a ladybird to 'fall' off or leave a plant depends on leaf morphology or plant structure (Kareiva & Sahakian, 1990; Grevstad & Klepelka, 1992; Frazer & McGregor, 1994). In contrast, the studies of Messina *et al.* (1997), Clark & Messina (1998) and Messina & Hanks (1998) indicate that the Russian wheat aphid, *Diuraphis noxia,* is at greater risk of predation from *Propylea quatuordecimpunctata* when feeding on Indian rice grass than on crested wheat grass mainly due to differences in the availability of refuges for the aphid on the two grasses. Proportionately more aphids feed in exposed positions on Indian rice grass than on crested wheat grass, especially

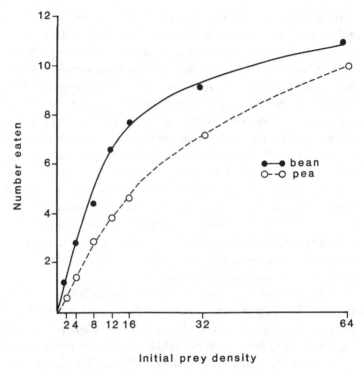

Fig. 5.30. The functional response of third instar larvae of *Coccinella septem-punctata* to increasing prey density on bean and pea plants of the same surface area. (After Carter *et al.*, 1984.)

when the aphid is abundant. That is, in this case the difference in predation risk is a consequence of the aphid's tendency to feed more frequently in relatively concealed positions on crested wheat grass rather than being due to differences in the searching behaviour of the predator on the two grasses.

SURVIVAL

If adults are successful in monitoring patch quality and as predicted by optimal foraging theory (Kindlmann & Dixon, 1993) lay a few eggs early in the development of a patch of prey, and the larvae are good at staying in the patch and harvesting the aphids, then larval survival should be high. That is, larvae in signalling their presence to ovipositing adults would appear to be creating a situation that is closer to contest than to a scramble for resources. However, the field data on survival of the immature stages of ladybirds indicate that it is very low (Table 6.2; p. 148),

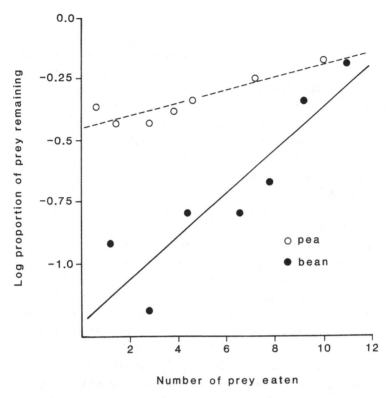

Fig. 5.31. Linear transformation of the functional response of third instar larvae of *Coccinella septempunctata* to increasing prey density on bean and pea plants. (After Carter *et al.*, 1984.)

and that much of the mortality can be attributed to cannibalism. Not exceeding the optimum number of eggs per patch, and thus maximizing the number of larvae maturing, depends on adults being able to assess the number of larvae each patch can sustain and the number already present. In addition, factors other than larval abundance determine the future trend in aphid abundance within a patch. That is, the future quality of a patch is very uncertain. In these circumstances possibly the best strategy is for adults to continue ovipositing until cues indicate that it is highly probable the eggs will be eaten. This is likely to result in the optimum number of eggs per patch being exceeded. In view of the uncertainty about future patch quality this is possibly the best strategy, with the fittest adults those that lay most of the early eggs. This aspect of ladybird ecology is in need of further study.

LADYBIRD ABUNDANCE

In 1925 Blattny suggested that the alternation of years of overpopu-
lation of aphids with years of lower aphid abundance is caused by the
action of aphidophagous insects. In years when aphids are abundant,
ladybirds and other natural enemies have suitable conditions for multiplication and the next year the large number of enemies can keep aphids
scarce. Because numbers of prey are reduced in this way, the population
increase of aphidophagous insects is inhibited in the second year and the
third year will again be favourable for aphids. That is, if the ratio of aphids
to their enemies is favourable for the latter, then aphid build-up is prevented. In contrast, if aphids are not subject to heavy natural enemy-
inflicted mortality early in a year their populations quickly explode. The
fact that many aphids tend to fluctuate in abundance from year to year
tends to support the idea, and several authors see it as the way aphidophagous insects operate in determining the abundance of aphids (Blattny,
1925; Hodek, 1967; Hagen & van den Bosch, 1968). It is likely that large
numbers of ladybirds emerging from hibernation could affect the establishment of aphid colonies in spring, especially as at that time these colonies consist of only a few aphids. However, it should not be forgotten that
the rate of increase of aphids is frequently restrained by intraspecific competition for resources (Dixon & Kindlmann, 1998) and the fitness of ladybirds is dependent on their locating patches of nursery prey that will
sustain the development of their larvae.

If one accepts that aphidophagous ladybirds forage as outlined
above then they are unlikely to regulate aphid abundance. Population
studies of tree-dwelling aphids tend to support this contention as the
delayed density-dependent mortality, so characteristic of classical predator–prey dynamics, is absent from these systems. Nevertheless, the abundance of aphids fluctuates from year to year, but this is attributed to
intraspecific competition between the aphids (Sequeira & Dixon, 1997;
Dixon & Kindlmann, 1999). In the absence of long-term studies on the
population dynamics of predatory ladybirds, however, their precise role
in determining aphid abundance will remain debatable.

One data set is helpful in resolving this debate. Heathcote (1970,
1978) meticulously recorded both the abundance of cereal aphids and the
numbers of hibernating ladybirds over a period of 15 years. Over the same
period some 60% of the adjacent land was planted with cereals, the
aphids on which were an important source of food for generalist aphidophagous ladybirds like A. bipunctata and C. septempunctata. Therefore, it
is reasonable to assume that most of the ladybirds produced each year in

this area are likely to have come from cereal fields. Ladybird swarms certainly coincide with outbreaks of cereal aphids (Carter *et al.*, 1982). In England in 1976, 280 000 *C. septempunctata* matured per hectare of wheat after feeding on cereal aphids, which nevertheless achieved a peak population of 200 million per hectare. The estimated 24 000 million *C. septempunctata* produced in the county of Norfolk combined with some produced in other counties gave rise to the ladybird 'plague' that drove people off the beaches in parts of East Anglia in that year (Dixon, 1998). Although large numbers of these ladybirds overwintered successfully the peak numbers of aphids on cereals in 1977 was higher than in 1976 (Carter *et al.*, 1982), which does not conform to the Blattny pattern.

If ladybirds regulate aphid abundance then in years when aphids are abundant they should eat a greater percentage of the aphids than in years when they are uncommon. That is, proportionally more ladybirds should develop in cereal fields and overwinter in years when aphids are abundant than when they are uncommon. The data collected by Heathcote were analysed by Carter *et al.* (1982) with this in mind (Fig. 5.32). The slope of the relationship between the logarithm of the numbers of hibernating ladybirds and the logarithm of the number of cereal aphids recorded the previous summer ($b = 0.93$) is not significantly different from 1.0, which supports the view that the abundance of aphids determines the abundance of ladybirds and not the reverse.

In the absence of detailed quantitative data prediction of future ladybird abundance is hazardous. In 1990 ladybirds emerged from hibernation early and species like *C. septempunctata* began mating and ovipositing a month earlier than average. The reason given for why these good early signs of ladybird well-being did not result in the expected population explosion of ladybirds was that they were killed by their natural enemies (Majerus & Majerus, 1996). Although there is no doubting that certain aphids, such as the nettle aphid, were abundant that year, the numbers of aphids on cereals in south-east England in 1990 were among the lowest recorded during the period 1977 to 1992 (Maudsley, 1993). Therefore, it is more likely that the generally low numbers of aphids on cereals, where most ladybirds develop, accounts for the low numbers of ladybirds in 1990.

In conclusion, the foraging behaviour of ladybirds, especially that of larvae, is well studied. As ladybirds seek out high-density patches of prey for breeding their functional response to patches containing low numbers of prey would appear to be irrelevant. This is particularly true of larvae, which are confined to the patch of prey selected by their mother. Those ladybirds that feed on prey that is well

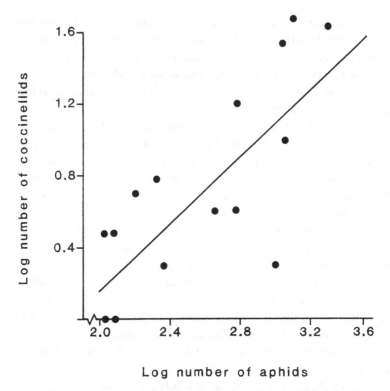

Fig. 5.32. Relationship between the number of coccinellids overwintering and the numbers of cereal aphids caught on sticky traps the previous summer at Brooms Barn, U.K., 1961–75. (After Heathcote, 1978.)

defended chemically or physically, or ant-attended, tend to be more prey specific. The more polyphagous ladybirds each oviposit in patches of a set of nursery prey characteristic of the preferred habitat of that particular ladybird. The ranking of prey is likely to be dependent on several factors, some of which vary in time. As predicted larvae do not show switching between prey dependent on their relative abundance, whereas adults of some species do.

The fitness of ladybirds is dependent on their ability to locate and oviposit in high-quality patches of prey. There are indications that adults use visual and olfactory cues to find first the appropriate habitat and then a patch of prey within that habitat. Optimal foraging theory predicts that aphidophagous ladybirds should lay a few eggs early in the development of a patch of nursery prey. In assessing the quality of a patch they do not appear to use cues associated with the age structure of the aphid colony or phenological age of the plant, but respond positively to high aphid abundance and negatively to the presence of conspecific larvae. This results in them mainly ovipositing early in the development of a patch of prey. The tracks of larvae contain a pheromone that deters adults from ovipositing. This response is adaptive because non-sib egg cannibalism is a major cause of egg mortality. The laying of eggs in clusters by aphidophagous ladybirds may also be adaptive as it possibly facilitates social feeding by first instar larvae, and thus increases their chance of surviving to the next instar. Larvae, like adults, show area-restricted search after eating prey and on encountering cues indicating the presence of prey. Hunger-induced changes in search behaviour serve to keep them in a patch and enable them to differentially exploit the clumping of prey within a patch and so maximize their energy intake. The hairiness or waxiness of plants can reduce the effectiveness of ladybirds in capturing prey, whereas a refuge for prey can increase their effectiveness especially at high prey densities.

The very high mortality of the immature stages of ladybirds in the field indicates that adults lay considerably more eggs than the optimum number of eggs in each patch. This is possibly the best strategy when there is great uncertainty about the future quality of a patch. If foraging by adult aphidophagous ladybirds is nevertheless close to optimal then they are unlikely to regulate aphid abundance. A long-term census of cereal aphid and ladybird abundance tends to support this contention.

6

Cannibalism

INTRODUCTION

Cannibalism refers to the act of killing and eating either all or part of an individual of the same species. It does not include cases in which individuals eat a conspecific that is already dead, which is conspecific scavenging. Traditionally cannibalism has been regarded as abnormal (Elgar & Crespi, 1992) as is well illustrated by a note published in *Science Gossip* in 1889. This records that J.W. Slater witnessed larvae of *Adalia decempunctata* engaged in the *reprehensible* habit of attacking and eating pupae of its own species, which he concluded seriously interferes with the multiplication of ladybirds and limits their usefulness as destroyers of aphids. However, cannibalism is widely recorded in ladybirds and is now regarded as part of their normal foraging behaviour.

Generally cannibalism is associated with an asymmetry between cannibal and victim. The victim is usually at a vulnerable stage in its development (Agarwala & Dixon, 1992; Dong & Polis, 1992; Stevens, 1992), i.e. in the egg or pupal stage, or is smaller or about to moult or pupate. Not surprisingly, given the potential risks, cannibalism is uncommon among individuals at the same stage of development. It also tends to occur most frequently when prey is scarce or predators are abundant. This leads to the prediction that the incidence of cannibalism should be highest in the egg and pupal stages, and in the larval stages greatest in the fourth instar when prey is likely to be scarce. Life table studies done on field populations support this prediction (Fig. 6.1; Yasuda & Shinya, 1997).

THEORY

The ecology and dynamics of cannibalism have been reviewed extensively by Fox (1975), Dong & Polis (1992) and Elgar & Crespi (1992). It

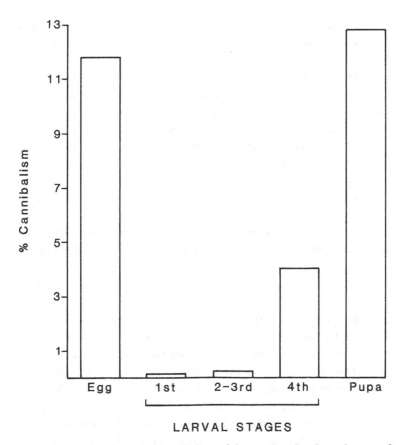

Fig. 6.1. The percentage cannibalism of the egg, larval and pupal stages of *Harmonia axyridis* observed in the field. (After Yasuda & Shinya, 1997.)

has been viewed as the redeployment of resources. A parent that produces a clutch that is partly consumed by her offspring is providing them with nutrition via what have been referred to as 'trophic eggs' and 'trophic larvae'. This would only appear to be a viable strategy if the parent is constrained in the size of the eggs she can lay. If not then the alternative of producing fewer but larger eggs would appear the more viable strategy, especially bearing in mind that providing adult size remains the same then larger eggs should also result in a shorter developmental time, which is particularly important in aphidophagous ladybirds (p. 102). That is, the offspring from larger eggs would have a double benefit.

In terms of optimal foraging cannibalism simply involves the addition of another item into an organism's diet. However, an individual that attempts to kill and eat individuals of roughly the same size and predatory ability runs a high risk of being damaged or falling prey to their potential victims. In addition, they run a risk of infection from parasites, viruses and diseases, a cost that has been largely ignored but should be considered in view of the burgeoning literature (e.g. Majerus & Hurst, 1997) on the widespread incidence of male-killer diseases in ladybirds.

Another more intriguing cost is the potential loss of fitness when a cannibal consumes a genetically related victim. Clearly the loss of fitness will depend on the degree of relatedness. If they are unrelated it is referred to as heterocannibalism and if they share a parent, sibling cannibalism. In the case where an adult eats its offspring it is referred to as filial or parental cannibalism. If they eat their kin, which by definition have similar genotypes, they are reducing the transfer of shared genes to the next generation and adversely affecting their inclusive fitness. Therefore, if principally a mechanism for acquiring food then the cost associated with the potential loss of inclusive fitness suggests that they should avoid eating kin. Thus the laying of eggs in clusters by aphidophagous ladybirds, which greatly increases the potential for sibling cannibalism, is all the more puzzling.

In assessing the cost associated with cannibalism it should not be assumed that they are always greater than those associated with foraging for prey. As in cannibalism the risks associated with foraging for prey depend on the vulnerability of the prey. Relatively large prey frequently damage or incapacitate predators, and small ladybird larvae kicked off a plant or heavily waxed by aphids may starve to death (Dixon, 1958). Studies on the interactions that occur when ladybird larvae encounter one another, similar to that done by Duelli (1981) on lacewing larvae and to the above and other studies done on the interactions observed between ladybird larvae and their prey (Klingauf, 1967; Wratten, 1976; Stadler, 1991), help in defining when cannibalism is advantageous.

The level of cannibalism is also likely to vary between individuals. In *Tribolium* there are strains that show high and low levels of cannibalism. This is genetically controlled with environmental factors influencing the expression of the trait. It appears that in this beetle larval cannibalism of eggs is controlled by many genes whereas the between-beetle strain differences in adult cannibalism of eggs may be due to as few as two genes. Larval and adult cannibalism of eggs have a high genetic correlation (Stevens, 1992). If also true of ladybirds then selection will favour an optimum level of cannibalism in a given environment. That is, cannibalism should be selected against in the absence of benefit to the individual but when such a benefit exists, it should evolve to a single environmentally dependent optimum. There is no empirical evidence that the level of cannibalism varies within a species, possibly because nobody has looked for it. However, researchers often state that some species are more difficult to rear collectively because they show higher levels of cannibalism than other species (Hemptinne, personal communication). It is likely that the level of cannibalism and the tendency to attack individuals of other species of ladybird is linked (p. 180). If this is the case then the prediction is that top intraguild predators are likely to show high levels of cannibalism. This would again argue for the level of cannibalism being viewed as a consequence of the factors determining diet breadth in a particular species. Cannibalism, although recorded (Nadel & Biron, 1964; Kehat, 1968), is less frequently commented on in studies of coccidophagous than of aphidophagous species of ladybird. This is possibly a consequence of the way coccidophagous species forage for prey – the eggs and young larvae are often hidden beneath their prey and therefore at less risk of being encountered by larger larvae, and because the abundance of their prey is unlikely to decline dramatically before they mature, the large larvae are less likely to be short of food.

CANNIBALISM BY ADULTS

As stated in Chapter 5 (p. 106) adult ladybirds avoid laying eggs in patches of prey where larvae are present. This could be seen as a case of minimizing μ/g, where μ is the probability of death per unit time and g the growth rate per unit time (Werner & Gilliam, 1984; Gilliam & Fraser, 1987; Gilliam, 1990). That is, an adult should choose a patch i for ovipositing in, if $\mu_i/g_i < \mu/g$, where μ/g is the average for all patches in the habitat. However, adults are unlikely to have the sensory capability for such an assessment. As indicated in Chapter 5 (p. 107) it is more likely they respond to simple external and internal cues that enable them to forage as if minimizing μ/g. As well as being adaptive in that their response to conspecific larvae reduces the probability of their eggs being eaten it also reduces the probability of adults being present along with their larval stages. This contention is well supported by the few life tables that have been published (Yasuda & Shinya, 1997). That is, ladybirds do not regularly coexist in space in age- structured populations, a feature that is more characteristic of invertebrate and vertebrate predators in aquatic systems.

Conspecific eggs appear to be a rich source of energy for both larval and adult ladybirds (Koide, 1962; Pienkowski, 1965; Dimetry, 1974; Kawai, 1978; Takahashi, 1987; Agarwala, 1991; Agarwala & Dixon, 1992). This is well illustrated by the two-spot ladybird, *Adalia bipunctata*, which can complete its development and lay eggs on a diet of conspecific eggs, although they may suffer a higher mortality and delayed development

(Koide, 1962; Kehat, 1968; Dimetry, 1974; Duelli, 1981). As the studies of Geyer (1947a, b) and Agarwala & Dixon (1992), on a coccid- and an aphid-feeding ladybird respectively, indicate that a diet of eggs is better than one of their usual prey, the supposed costs may be more a consequence of failing to maintain an adequate supply of eggs/food than to any detrimental effects a diet of eggs may have on survival and development. However, the cost in loss of inclusive fitness if adults eat closely related kin indicates that they should avoid eating their own eggs. When given a choice of eating a cluster of her own eggs or those of another female, the two-spot ladybird, A. bipunctata, tends to eat those of the other female first (Agarwala & Dixon, 1993a). This has also been observed in Coccinella hieroglyphica (Hippa et al., 1982). That is, in these species, as theory predicts, egg cannibalism by adults is likely to involve mainly eating the eggs of non-kin.

Summarizing, in the field cannibalism of larval stages by adult ladybirds is likely to be relatively uncommon, mainly because of the tendency of adults to leave patches of prey that contain conspecific larvae. Egg cannibalism, however, is likely to be more common but as predicted by theory there is empirical evidence to indicate they avoid eating their own eggs.

CANNIBALISM BY LARVAE

Unlike adult ladybirds their immature stages are usually confined to a patch of prey (p. 94). That is, their potential scope for choosing among sites differing in prey abundance and predator risk is very limited. Therefore the frequency of encounters between immatures is likely to be very high. As with adults there appear to be no physiological costs associated with cannibalism compared to eating their usual prey. Therefore, foraging theory predicts that they should attack and eat a food item if the benefit (B) to cost ratio (B/C_p), where C_p denotes the foraging cost associated with a particular prey item, is higher than the threshold value (B/C) for the patch. However, there may be greater costs associated with pursuing and overpowering conspecific larvae and/or penetrating the exoskeleton of pupae than in capturing prey, which would reduce the value B/C_c, where C_c denotes the foraging costs for conspecifics. When B/C_c is greater than B/C_p then cannibalism is advantageous. A major factor determining cannibalism is conspecific vulnerability. Therefore, it is not surprising that eggs, small and starving larvae, moulting larvae and pupae are more frequently the victims of cannibalism than are large well-fed larvae. This asymmetry in size or vigour between cannibal and victim is a marked feature of cannibalism. In addition, scarcity of prey through hunger, which affects vulnerability and voracity, is also likely to promote cannibalism (cf. Fig. 6.1). That is, size, developmental stage, and hunger are likely key determinants of relative vulnerability and when $B/C_c > B/C_p$ cannibalism should occur. Cannibalism also confers competitive benefits on cannibals and therefore conspecifics may still be eaten if $B/C_c < B/C_p$. If B/C_c is consistently less than the average B/C_p then cannibalism may have evolved primarily as a mechanism that reduces competition (Dong & Polis, 1992). In addition, if inclusive fitness is important then larvae should avoid eating kin. There is some evidence for this from the laboratory study of Agarwala & Dixon (1993a), which shows that not only are the adult females of A. bipunctata more reluctant to eat their own eggs than those of other females, but the larvae of this species are similarly more reluctant to eat sibs, be they eggs or larvae, than non-sibs.

Fitness

As many aphidophagous ladybirds mate more than once and tend to lay clusters of eggs the individuals in a cluster can be siblings or half-siblings depending on whether they share one or both parents. The adaptiveness of sibling cannibalism in view of the potential loss of fitness associated with eating kin has been considered by Osawa (1992a). He did this by determining the increase in the probability of survival from first to second instar of cannibals relative to that of their victims in Harmonia axyridis, at two aphid population densities. In this context the fitness of a larva can be designated p, the probability of survival to the next instar, and that of a cannibal is increased by p^* on consuming one victim, so its new fitness is $(p + p^*)$. Hamilton (1964) showed that the inclusive fitness of an individual will be increased, and the selfish trait (cannibalism) selected for if $-K < 1/r$, where r is the coefficient of relatedness. In this equation K represents the change in fitness of the larva eaten divided by the gain in fitness of the cannibal.

$$\Delta \text{ fitness of larva eaten} = o - p = -p$$
$$\Delta \text{ fitness of cannibal} = (p + p^*) - p = p^*$$
$$K = -p/p^*$$

In the case of siblings $r = 0.5$ and half-sibs 0.25. For cannibalism to be selected for $p/p^* < 2$ for sibs and 4 for half-sibs.

The results presented in Osawa's (1992a) paper and Table 6.1 indicate that cannibalism is generally adaptive when aphids are scarce. Interestingly, when aphids are abundant cannibalism is advantageous if

Table 6.1. *The advantage in terms of fitness to a first instar larva of* Harmonia axyridis *of eating 0, 1 or 3 sibling larvae when aphid abundance is low or high*

Aphid abundance	Number of larvae eaten	Probability of survival first–second instar	Fitness of		p/p^*	Cannibalism advantageous?
			Larva(e) eaten, $-p$	Cannibal, p^*		
Low	0	0.124				
	1	0.404	−0.124	0.280	0.44	yes
	3	0.778	−0.372	0.654	0.57	yes
High	0	0.343				
	1	0.398	−0.343	0.055	6.20	no
	3	0.886	−1.029	0.543	1.90	yes

Note:
After Osawa, 1992a

the cannibal eats three sibs but not if it eats only one. This appears to present a very serious problem in terms of decision-making for larvae as it would appear they have to eat either three sibs or none! However, this may be a consequence of the very slight improvement in survival recorded when one sib is eaten and aphids are abundant. If this result is anomalous and is in reality closer to 0.6 as predicted by the trend in the increase in probability of survival observed when aphids are scarce then the problem is resolved as p/p^* associated with eating one sib becomes 1.3 and cannibalism is advantageous. That is, even when aphids are abundant it is likely that larval cannibalism is advantageous, mainly because it increases the probability of survival so markedly.

Currency

The advantages of cannibalism are viewed in terms of net energy intake, which raises questions about how insects might assess the costs and benefits. Hunger is an important determinant of foraging behaviour in ladybirds (p. 122). Therefore, another way of viewing cannibalism, which more fully recognizes that such predators operate under severe physiological and sensory constraints, is to view cannibalism more in terms of the relative risk of death from starvation or being eaten (p. 84). When well fed, a larva that encounters something with a waxy covering similar to its own, be it egg, larva or pupa, should be cautious as it could become the victim. If hungry then to attack a conspecific is still risky, but the risk is now relative to that of starving to death. That is, they do not recognize eggs as such but as a conspecific, which might overpower and kill them. If this is correct then it accounts for why well-fed but hungry larvae are so reluctant to attack and eat conspecific eggs, which appear to be a very easy source of high quality food. The response to larvae is likely to be similar, but the relative mobility and vigour of the larva encountered are also likely to be important in determining the outcome.

If ladybirds are successful in assessing patch quality then at the time of egg hatch competition for food is unlikely to be important as aphids should then be increasing in abundance. Therefore, if sib cannibalism serves mainly to increase the proportion surviving to the second instar, and survival at this stage is generally very low, then it does raise the question: why not lay fewer and larger eggs?

As with so many processes it is important to ask whether ladybirds are likely to have the sensory ability necessary to asses the benefits and costs associated with foraging. When presented with two kinds of prey that differ in quality and abundance then if the relative costs and benefits are paramount in determining the foraging strategy one would expect a predator to switch and take mainly the most profitable prey. There are few studies on this aspect of foraging behaviour in ladybirds but they are consistent in indicating that ladybird larvae do not switch between prey (p. 93). Similarly there is no indication that they forage for a particular size of prey even though the foraging costs associated with pursuing and subduing prey of different sizes vary greatly (Dixon, 1958). Accepting that it would be advantageous for larvae to switch between different species of prey or different size classes of prey, but that they do not, might indicate that they are constrained by their sensory capability. If they do not have the level of omniscience to resolve this foraging problem what mechanism(s) are likely to govern the incidence of cannibalism in ladybirds?

Field observations (Yasuda & Shinya, 1997) indicate that ladybird larvae are frequently faced with a shortage of prey, especially in the later stages of their development, and under such circumstances cannibalism is advantageous as it results in at least some larvae surviving to reach maturity. As in foraging for prey cannibalism is not without risks. If not able to assess the relative risks, and larvae do appear to pursue all prey irrespective of their relative size, then to attack all animals encountered irrespective of their size is possibly the best strategy. The asymmetry of each such encounter determines the outcome irrespective of whether the potential victim is prey or a conspecific.

If this scenario is correct then the incidence of cannibalism is simply a consequence of the frequency of encounters between conspecifics and their relative vulnerabilty. That is, cannibalistic attack rates are determined by the probability of encountering vulnerable individuals. The advantage of an individual-based model of cannibalism is that it does not assume, as in population-state models, that all individuals are approximately equal and that there are stable emergent properties independent of individual variation.

MODEL

In the field it is very unusual for all the different stages of development of a ladybird to co-occur in the same patch of prey. There is a tendency for them to occur in a definite sequence: adults, eggs, larvae, pupae, adults, with a slight overlap between each stage. That is, asymmetry is most striking at the interphases between the developmental stages, in particular eggs/larvae and larvae/pupae.

Assuming that hungry ladybirds have a constant consumption rate and the proportion of conspecifics in their diet is proportional to their relative abundance then the number eaten is:

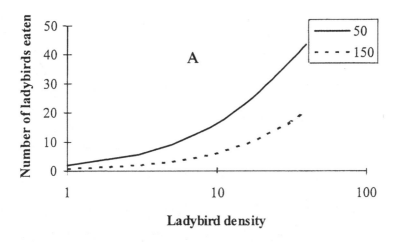

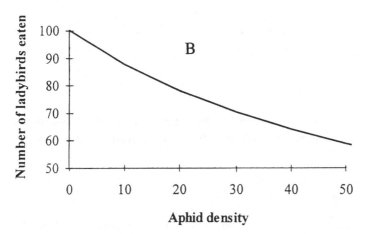

Fig. 6.2. The predicted (A) increase in cannibalism with increase in predator density at two levels of prey density, 50 and 150, and (B) the decrease in cannibalism with increase in aphid density when predator density is kept constant.

$$f(x, y) = ay/(x + y)$$

where x is the number of prey, y is the number of ladybirds and a is a scaling constant. This simple relationship predicts that if abundance of prey is kept constant the incidence of cannibalism increases with increase in predator abundance. At a higher density of prey there is a similar relationship but the rate of increase in the incidence of cannibalism with increase in predator abundance is less (Fig. 6.2(A)). Alternatively if the number of predators is kept constant and that of the prey is varied the model predicts that the incidence of cannibalism decreases with

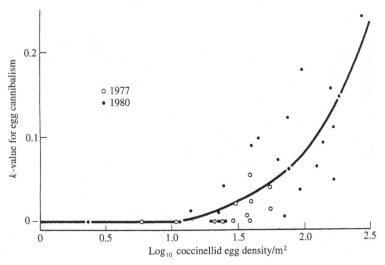

Fig. 6.3. The relationship between egg cannibalism and the number of eggs per unit area of lime foliage in *Adalia bipunctata* in the field. (After Mills, 1982*a*.)

increase in the abundance of prey (Fig. 6.2(B)). That is, cannibalism should be density dependent and potentially capable of regulating the numbers of predators.

EMPIRICAL DATA

The prediction that cannibalism is capable of regulating the numbers of predators is supported specifically for ladybirds by the results of Mills (1982*a*) and Osawa (1993), and has been shown for other organisms (Fox, 1975; Polis, 1981; Dong & Polis, 1992). Mills's (1982*a*) classic study of the relationship between egg cannibalism and egg density in *A. bipunctata* in the field revealed that cannibalism can act as a strong density-dependent mortality factor (Fig. 6.3). A few laboratory studies complement the field studies in addressing problems such as the effect of prey density which are difficult to manipulate and control for in field experiments. More studies of the type done by Duelli (1981) on the larvae of another group of aphidophagous predators, the lacewings, are needed. Because of their mobility and tendency to disperse throughout a patch the fate of this stage in the development of ladybirds is difficult to monitor. The few life tables there are of field populations of ladybirds appear to indicate that mortality due to cannibalism increases with

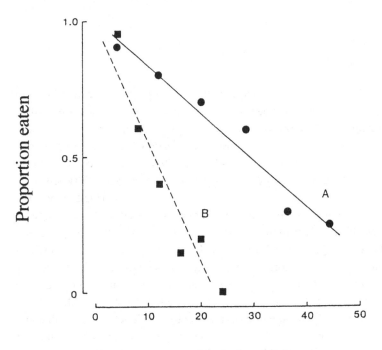

Number of aphids

Fig. 6.4. The incidence of cannibalism in the laboratory of clutches of eggs (A, ●——●) and larvae (B, ■ - - - ■) of *Adalia bipunctata* in relation to aphid abundance. (After Agarwala & Dixon, 1992.)

instar (Fig. 6.1; Yasuda & Shinya, 1997). This is most likely a consequence of the number of predators relative to those of the prey changing in time. Early on aphids are relatively abundant but late on they become relatively scarce and as a consequence both the frequency of encounters between hungry larvae and the incidence of cannibalism increase. The predicted increase in larval survival when prey density is high has been reported in several field studies (Yasuda &Shinya, 1997). However, it is not always clear whether predator or prey density, or an interaction between them, determined the outcome. A laboratory study on *A. bipunctata*, however, indicates that prey density is important in determining larval survival as cannibalism decreases as prey population increases (Fig. 6.4; Agarwala & Dixon, 1992)

Eggs

The egg is a very vulnerable stage, with percentage cannibalism in the field ranging between 6 and 60% (Mills, 1982a; Osawa, 1989). Mills attributes little of this to sib cannibalism in *A. bipunctata*, whereas Osawa indicates that 25% of the egg losses in *H. axyridis* can be attributed to this cause. This difference in the incidence of sib cannibalism could be due to male-killer diseases, which are known to vary in incidence between populations both within and between species (Majerus & Hurst, 1997). The high incidence of sib cannibalism of eggs observed in the laboratory is possibly due to several factors. One is the rearing conditions, in particular the tendency to feed adults once a day. As the quality of the aphids supplied tends to deteriorate very quickly it is likely the adults eat relatively few aphids over a short period each day, and as a consequence eggs are matured in their gonads over a longer period than in the field. This may account for the considerable delay between the hatching of the first and the last egg in a cluster, often observed in the laboratory, which contrasts with the synchronized hatching observed in the field. In the laboratory there is no correlation between the incidence of cannibalism and cluster size (Kaddou, 1960; Dixon & Guo, 1993) but a strong correlation with the delay in time between the hatching of the first and the last eggs in a cluster (Fig. 6.5; Kaddou, 1960). In addition it may also be attributable to inbreeding, which in both *A. bipunctata* and *Propylea quatuordecimpunctata* results in an increase in the proportion of eggs that fail to hatch (Hurst *et al.*, 1996; Morjan *et al.*, 1999) and are therefore likely to be eaten by the larvae that do hatch.

Because eggs are so vulnerable to cannibalism one would expect aphidophagous ladybirds, which tend to lay their eggs in clusters in exposed locations, to spend relatively less time in the egg stage than coccidophagous species, which tend to lay their eggs singly beneath prey. In absolute terms they do, e.g. at 25 °C coccidophagous species spend twice as long in the egg stage as aphidophagous species (p. 69). In addition the data in the literature indicate that aphidophagous species also spend less time in the egg stage relative to that spent in the pupal stages than do coccidophagous species (Fig. 6.6). In terms of a trade-off it would be interesting to know if the precocious hatching has an effect on other life history parameters. It is unlikely that it predisposes the hatching larvae of aphidophagous species to be more vulnerable to starvation as implied by Majerus (1998). The size and reserves of neonate larvae is most likely a simple function of egg size.

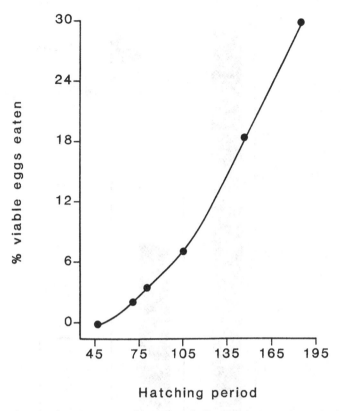

Fig. 6.5. The percentage sib egg cannibalism in clusters of eggs laid by *Hippodamia quinquesignata* in the laboratory in relation to the time (in days) between the first and the last egg hatching. (After Kaddou, 1960.)

Because of the risk from non-sibling cannibalism eggs should not be laid at a time and a place where they are likely to be easily located by larvae. The optimum time for oviposition in *H. axyridis* appears to be early in the developments of a patch, which is the time when very few larvae are present (Fig. 6.7(A); Osawa, 1989). As the behaviour of the larvae leads them to concentrate their foraging in the vicinity of prey, eggs should be laid some distance from prey. However, if too far from aphids the larvae are unlikely to survive. Several authors have recorded that ladybirds often lay their eggs some distance from aphids (Banks, 1954, 1957; Dixon, 1959; Osawa, 1989), and Osawa (1989) showed that 56% of the eggs laid either 10 cm or closer to aphids were eaten by non-sibling larvae, in contrast to only 33% of those laid further than 10 cm away from aphids (Fig. 6.7(B)). The poor dispersal powers of newly hatched first instar larvae, however, place a severe constraint on this as a means of avoiding cannibalism. In addition it is difficult to conceive of how adult ladybirds might accurately

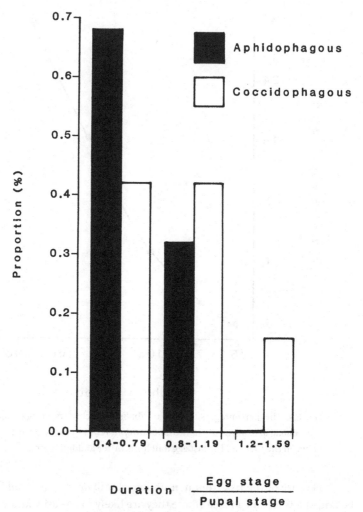

Fig. 6.6. The proportion of aphidophagous and coccidophagous species of ladybird in which the duration of their egg stage relative to that of the pupal stage is 0.4–0.79, 0.8–1.19 or 1.2–1.59. (Data from: Aguilera, 1995; Ahmad, 1970; Bain *et al.*, 1994; Benham & Muggleton, 1970; Booth *et al.*, 1995; Brettell, 1964; Brown, 1972; Butler, 1982; Butler & Dickerson, 1972; Campbell *et al.*, 1980; Chakrabarti *et al.*, 1988, 1995; Chazeau *et al.*, 1991; Correjo *et al.*, 1991; de Fluiter, 1939; Geyer, 1947*a*; Ghani & Ahmad, 1966; Greathead & Pope, 1977; Gurney & Hussey, 1970; Hecht, 1936; Jalali & Singh, 1989*b*; Kawauchi, 1983; Meyerdirk, 1983; Michels & Bateman, 1986; Moursi & Kamal, 1946; Nadel & Biron, 1964; Naranjo *et al.*, 1990; Okrouhlá *et al.*, 1983; Pantyukhov, 1968; Raghunath & Rao, 1980; Schanderl *et al.*, 1985; Sharma *et al.*, 1990; Simpson & Burkhardt, 1960; Toccafondi *et al.*, 1991; Varma *et al.*, 1993; Zhao & Wang, 1987.)

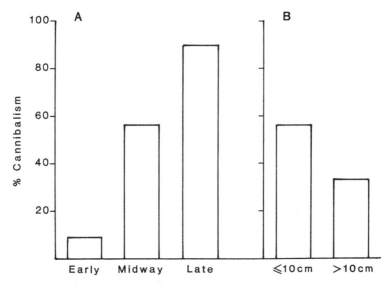

Fig. 6.7. The percentage egg cannibalism in clusters of eggs of *Harmonia axyridis* (A) laid early, midway and late in the development of a patch of prey, and (B) close to (≤10 cm) and more distant from aphids (>10 cm). (After Osawa, 1989.)

assess the distance to the nearest aggregate of aphids. Before speculating further it is necessary to establish that proportionally more of the eggs are deposited at a distance from aphids than one would expect by chance.

As indicated in the development of the model presented on p. 139 the numbers of prey relative to the number of larvae is also likely to affect the incidence of cannibalism. Osawa's (1992a) study on *H. axyridis* also reveals that the incidence of non-sib egg cannibalism increases from 9% early in the development of an aphid patch, through 56% in the middle period, to 90% late on in the existence of the patch (Fig. 6.7(A)). Although, as claimed by Osawa, the high incidence of cannibalism in the middle and late phases in the development of a patch is mainly due to the high density of *H. axyridis* larvae relative to aphid density, the large size and mobility of the larvae late on in the development of the patch is also likely to have been important. However, it does clearly indicate that ladybirds should avoid laying eggs late on in the development of a patch as they have little chance of surviving. The emprical and experimental data tends to support this prediction (Hemptinne *et al.*, 1992).

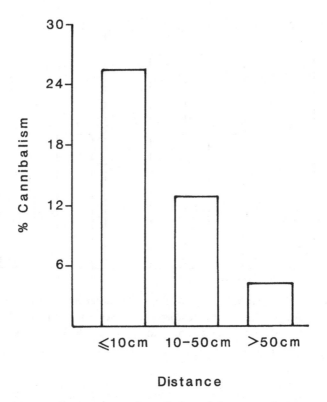

Fig. 6.8. The percentage of cannibalism of the pupae of *Harmonia axyridis* that pupate close to ($\leq$10 cm) and more distant (>10–50 cm; >50 cm) from aphids. (After Osawa, 1992*b*.)

Pupae

As with eggs the immobility of the prepupal and pupal stages makes them particularly vulnerable to cannibalism. It is not unusual to find the pupae of ladybirds at some considerable distance from the aphid colony they fed on as larvae. For example, *A. bipunctata* frequently pupates in very large numbers on the trunks of lime trees at some considerable distance from the leaves that support their aphid prey. The adaptive significance of this has been studied by Osawa (1992*b*). In the case of *H. axyridis* he showed that the closer it pupates to aphids the greater the risk of cannibalism (Fig. 6.8). Those that move off the host plant of their prey and pupate on adjacent vegetation experienced an even lower incidence of cannibalism. In addition those that pupate on adjacent host plants also suffered less parasitism. Therefore, as expected there are considerable advantages to pupating some distance from prey. Nevertheless, as 78% of the pupae of *H. axyridis* were found on the host plant of the prey, dispersal would appear to be risky (Osawa, 1992*b*), or difficult to achieve. As with the avoidance of egg cannibalism a rigorous study of the behaviour of prepupal larvae that might reduce the incidence of pupal cannibalism and predation is needed.

CANNIBALISM AS A MEANS OF HARVESTING PREY – THE ICEBOX HYPOTHESIS

Temporal changes in the abundance of aphids are dependent on many factors: natural enemies, weather, response of plant to aphid infestation, intraspecific competition and response of aphids to the presence of natural enemies. Therefore, the duration of time for which the aphids in a patch remain sufficiently abundant to support the development of ladybird larvae is very uncertain. As ladybirds take considerably longer to complete their development than aphids this uncertainty over the future availability of prey is important. In addition, as adults abandon their eggs they would appear to have little ability to control the distribution of resources between their offspring. However, the tendency of aphidophagous species to lay their eggs in clusters, whereas coccidophagous species tend to lay them individually, may indicate that the former are manipulating the supply of resources for their offspring and so maximizing their fitness. In this context the many more reports of high incidences of cannibalism in aphidophagous than coccidophagous ladybirds is interesting. Is this a consequence of the larvae of aphidophagous species more frequently encountering food shortages than coccidophagous species, or is cannibalism a means of harvesting prey?

Optimal foraging theory indicates that ladybirds should lay a few eggs early in the development of a patch of aphids. Empirical data indicates they do lay their eggs more frequently early than late on in the development of a patch (p. 107). If this happens successfully one would expect most larvae to complete their development, i.e. the larvae in producing a pheromone that deters adults from ovipositing are reducing the likelihood of a scramble for food, which would result in high levels of mortality. However, in view of the magnitude of the environmental uncertainty it is unlikely a parent can predict the optimal clutch size. In this case to use some of its offspring to increase its reproduction via others appears to be a more viable strategy. The few aphidophagous ladybird life tables in the literature clearly indicate very high levels of mortality (83–100%) of which a large proportion is attributed to cannibalism (Table 6.2). Thus, the larvae of aphidophagous ladybirds apparently regularly scramble for food and suffer high levels of cannibalism.

Laying eggs in clusters increases the chances that larval and pupal

Table 6.2. *The percentage mortality over the period from oviposition to adult emergence, recorded in the field, for six species of ladybird*

Ladybird	Plant/prey	Year	Egg to adult, percentage mortality	Source
Aiolocaria hexaspilota	*Juglans mandschuria/Gastrolina depressa*		99.6	Matsura (1976)
Coccinella septempunctata	*Hibiscus syridis/Aphis gossypii*	1993	100	Yasuda & Katsuhiro (1997)
		1994	95.4	
Coleomagilla maculata lengi	*Zea mays/Rhopalosphum maidis*	1977	97.1	Wright & Laing (1982)
		1978	95.4	
Harmonia axyridis	*Hibiscus syridis/Aphis gossypii*	1993	98.9	Hironori & Katsuhiro (1997)
		1994	92.9	
	Cyrnara scolymus/Capitophorus elaeagni	1987	99.2	Osawa (1992c)
	Seven plant/aphid combinations	1987	87–100	Osawa (1993)
		1988	83.9–100	
Hippodamia convergens	*Sorghum bicolor/Rhopalosiphum maidis* and *Schizaphis graminum*	1972	96.3–98.6	Kirby & Ehler (1977)
		1973	93.1–99.6	
Hippodamia tridecimpunctata tibialis	*Zea mays/Rhopalosiphum maidis*	1978	91.5	Wright & Laing (1982)

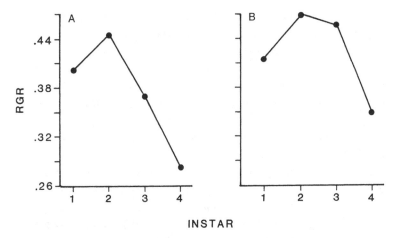

Fig. 6.9. The trends in relative growth rates of the larval instars of (A) *Semiadalia undecimnotata* and (B) *Adalia bipunctata*. (Data from: Ferran & Larroque, 1977; Ferran *et al*., 1984.)

cannibalism will involve eating kin. Initially the ladybird larvae are likely to find themselves surrounded by an abundance of young aphids on which they can feed; however, once aphid abundance starts to decline cannibalism may be the only way some of them will survive to maturity. That is, while the aphids are increasing in abundance the many small ladybird larvae each eat a few aphids, which overall may have little or no affect on the rate of increase of the aphids because the loss of aphids reduces the intraspecific competition between aphids. In addition, based on their relative growth rates, small larvae are better at converting aphid biomass into ladybird biomass (Fig. 6.9). Some of the biomass of ladybird larvae that develops during the early phase in the development of an aphid patch is later exploited by some of the larvae as a source of food. That is, victims essentially function as packages of live meat for their kin, a phenomenon dubbed the *icebox hypothesis* (Alexander, 1974). In the field there is a high incidence of cannibalism when larvae begin to pupate (Fig 6.1) which often ensures that a few large larvae survive to produce adults and breed.

Is it reasonable to suggest that the hunting strategy of aphidophagous ladybirds has similarities with that of African hunting dogs and wolves, with a few individuals in each cluster surviving to breed aided by the harvesting of prey by their kin? As indicated above there is a high probability that the cannibal and victim will be kin; however, this is unlikely to be the rule. The fact that larvae disperse throughout a patch reduces the potential for intrabrood altruism. Thus, the evolution of such a strategy in ladybirds poses problems. If not pure speculation then it might account for egg clustering and high larval mortality in aphidophagous ladybirds. However, the most likely interpretation of the low juvenile survival is that ladybirds lay more than the optimum number of eggs per patch, as the best strategy for adults is to lay as many eggs as early as possible in the development of a patch of prey (p. 102). If this is the case then cannibalism ensures that at least some of the offspring survive.

In conclusion, cannibalism is widely recorded in ladybirds, and is regarded as part of their normal foraging behaviour. Generally it is associated with an asymmetry in size or activity between cannibal and victim. Cannibalism of larvae by adults is likely to be rare, mainly because of the tendency of adults to leave patches of prey that contain conspecific larvae. Egg cannibalism by adults, however, is likely to be more common. As predicted by theory adults are reluctant to eat their own eggs. Although there is a potential loss of fitness associated with eating kin it nevertheless appears to be adaptive, mainly because it increases the probability of survival of larvae so markedly.

The empirical data indicates that the incidence of cannibalism is simply a consequence of the frequency of encounters between hungry conspecifics and their relative vulnerability. A simple model based on this idea predicts that if abundance of prey is kept constant the incidence of cannibalism increases with increase in predator abundance. If the number of predators is kept constant and that of the prey is increased the prediction is that the incidence of cannibalism decreases. The form of these relationships observed in field and laboratory studies conform very closely to these predictions.

Aphidophagous ladybirds lay their eggs in clusters in exposed locations, which makes them very vulnerable to cannibalism. Therefore it could be viewed as adaptive that they spend proportionally less time in the egg stage than coccidophagous ladybirds, whose single eggs laid beneath prey are less at risk of cannibalism. Although this is speculative, laying eggs in clusters and cannibalism may also be a means of harvesting prey. In this scheme the biomass of ladybird larvae that develops during the early phase in the development of a patch is later exploited by some of the larvae as a source of food. If the larvae are kin then this would be a case of intrabrood altruism. The evolution of such a strategy in ladybirds poses a problem as the larvae are unlikely to always be kin, although the clustering of eggs, so characteristic of aphidophagous ladybirds, increases the probability of cannibalism involving kin. It is more likely, however, that the high incidence of cannibalism in aphidophagous ladybirds is a consequence of their laying more than the optimum number of eggs per patch of prey.

7

Theory of predator–prey interactions

INTRODUCTION

All the insect natural enemies of aphids and coccids kill them for food. Therefore, in terms of the most common definition of predation (Taylor, 1984) they are all predators. However, they exploit their food in one of three ways.

Parasitoids

These are mainly wasps (Hymenoptera) and a few flies (Diptera), which insert an egg into an aphid or coccid where it hatches, grows and pupates, often remaining within the skin of its host throughout its immature development (Stary, 1970). Although the larvae of parasitoids may have to compete with other larvae they do not have to forage for food as they each develop within a unit of resource, which is adequate to sustain them throughout their development (Fig 7.1). These natural enemies are similar in size and only have a slightly longer developmental time than their hosts. Adult parasitoids are free-living and feed on nectar or body fluids of their host (e.g. Collins et al., 1981). In the earlier literature they were referred to as parasites (cf. Clausen, 1940) but because their larvae always kill their hosts they are now referred to as parasitoids. Their fitness depends on being able to locate and parasitize suitable hosts. That is, hosts that do not already contain parasitoid larvae and will live long enough to sustain the development of their offspring.

Predators

A few species of wasps of the family Sphecidae, such as *Bhopalum clavipes* and *B. coarctatum*, hunt for aphids, which they subdue by stinging

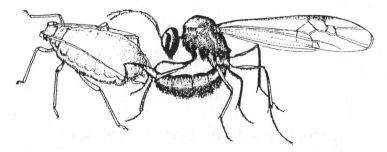

Fig. 7.1. Hymenopteran parasitoid ovipositing in an aphid.

and use to provision their nests. Once they have collected sufficient prey to sustain the development of a larva an egg is laid on the pile of aphids in the cell, which is then sealed off and the wasp proceeds to provision another cell (Fig. 7.2). The larvae of these wasps also do not have to forage for food as they are each surrounded by sufficient immobile prey to sustain their complete development. Although more than one aphid is required for development, the pursuing and subduing of the aphids is done by the adult wasp. In this case the natural enemy is both bigger and has a considerably longer developmental time than its aphid prey. In foraging for aphids the adult wasp is likely to visit many colonies of aphids and exploit those that give the greatest return per unit effort.

The last case is that of classical insect predators, such as aphidophagous ladybirds, which lay their eggs close to their prey. On hatching, however, the larvae have to pursue and subdue their prey. This is very different from the above where the larvae do not forage for food. Although many coccidophagous ladybirds lay a single egg under a scale insect and the larva achieves an advanced stage of development on that scale, nevertheless in most cases more than one scale is needed for complete development. That is, all the entomophagous ladybirds are classical insect predators in that their larvae require more than one prey item for their development and what is more important the larvae have to find and catch at least some of their prey. The fitness of the adult in this case is dependent on their locating patches of high concentrations of nursery prey, which last long enough to sustain the development of their larvae (Fig. 7.3).

THEORY

The Nicholson & Bailey approach to modelling predator–prey interactions has resulted in very complex models. It is likely that their very

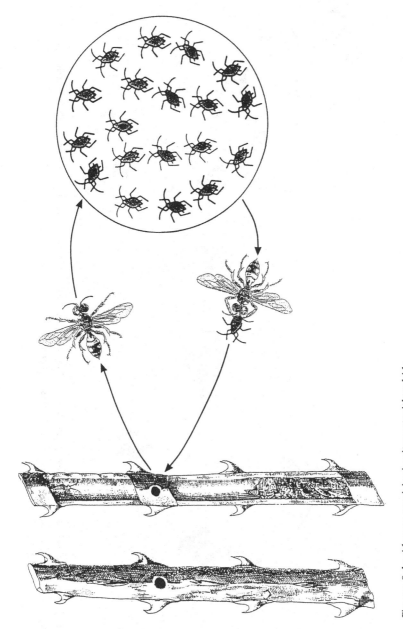

Fig. 7.2. Sphecid wasp provisioning its nest with aphids.

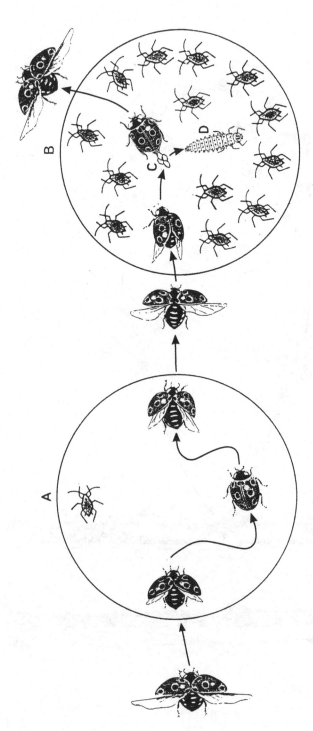

Fig. 7.3. Aphidophagous ladybirds quickly leave patches where aphids are scarce (A) but oviposit in patches where prey is abundant (B). The larvae (D) that hatch from the eggs (C) have to pursue and subdue the aphids they need for their development.

complexity has impeded the application of the theory of population dynamics as an aid in the practice of biological control. Other approaches, such as optimal foraging, give a better understanding of the selection pressures that are likely to have shaped ladybird ecology but in their current form are similarly unlikely to be of use to biological control practitioners. However, these optimal foraging studies have revealed the importance of developmental time. In addition the values of the other life history parameters are closely linked with developmental time (p. 80). That is, the easily measured developmental time is possibly a good measure of all the other attributes of a ladybird. If this is so a more minimalistic approach might better provide the answer to the question posed in Chapter 1: why is *Rodolia cardinalis* such an effective biological control agent?

Nicholson & Bailey

There has been a long and ongoing attempt to model the interaction between predators and prey. Insects make ideal subjects for such modelling because their generation times are characteristically short and many have relatively discrete generations inviting the use of difference equation models to describe population changes.

Central to understanding predator–prey dynamics is the relationship between the death rate imposed on the prey by the predators and the rate of increase or numerical response of the predator population. The favoured starting point has been the Nicholson–Bailey model (Nicholson, 1933; Nicholson & Bailey, 1935). This was formulated with insect parasitoids in mind. However, the assumptions on searching behaviour apply equally to predators: (1) the number of encounters with prey (N_e) by P_t predators is in direct proportion to host density N_t and (2) the N_e encounters are distributed randomly among the available prey. This gave the following equations:

$$N_{t+1} = \lambda \, N_t \, \exp(-aP_t) \qquad (7.1)$$

Number of prey in next generation	Net rate of increase of prey	Number of prey in this generation	Fraction of prey surviving	

$$P_{t+1} = N_t \, [1 - \exp(-aP_t)] \qquad (7.2)$$

Number of predators in next generation	Number of prey in this generation	Fraction of prey eaten	

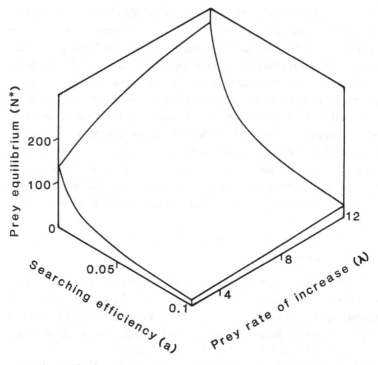

Fig. 7.4. The dependence of the equilibrium prey population (N^*) on both the predator searching efficiency (a) and the prey rate of increase (λ) (After Hassell, 1978.)

which revealed that when the population of predators and prey were in equilibrium, the searching efficiency of the predator (a) and the rate of increase (λ) of the prey determined the equilibrium density of the prey (Hassell, 1978; Fig. 7.4). This assumes there is a linear relationship between the number of prey killed and predator reproduction. Although valid for host–parasitoid interactions, the rate of increase of a predator is not a simple function of the number of prey consumed.

To overcome this problem Lawton *et al.* (1975) proposed that the number of prey attacked (N_a) should be framed in terms of prey (N) and predator (P) density.

$$N_a = aNT/[\,1 + aT_hN + bT_w\,(P - 1)\,] \qquad (7.3)$$

where a defines the rate of encounter between predators and prey, T_h is the handling time, b the encounter rate between predators and T_w the time wasted (mutual interference) on an encounter between predators. In those cases where predators do not search at random and prey is aggregated, a and b are functions of the relative distributions of predator and prey.

For a predator confronting a variety of prey densities equation 7.3 can be simplified to:

$$N_a = aNT/(1 + aT_hN) \qquad (7.4)$$

Holling's (1959, 1966) functional response, which defines the increase in number of prey attacked relative to prey density. The asymptote of the response results from the progressive decrease in the time available for foraging. As the number of prey captured increases, a greater proportion of the total time available (T) is spent handling rather than foraging for prey. As predators allocate some of the food they ingest for maintenance there is a threshold of food consumption below which growth and egg production ceases (Fig. 7.5). Therefore, both the growth rate (g) and fecundity (F) of a predator should be related to energy intake (I) by similar models:

$$g = \delta (I - c) \qquad (7.5)$$

$$F = \lambda (I - c) \qquad (7.6)$$

where the constant c is determined by the energy required for maintenance and the proportion of this ingested food that is assimilated, and δ and λ are proportions. If W is the increase in weight during development then the ratio W/g defines d, and the inverse of d, the developmental rate:

$$1/d = (\delta/W)(I - c) \qquad (7.7)$$

When food is scarce ladybirds often moult to the next instar at a lower body weight than when food is abundant (p. 137). This phenomenon can be described by an equation that is similar to equation 7.7:

$$1/d = \alpha (I - \beta) \qquad (7.8)$$

in which α and β have no simple biological interpretation.

The relationships between developmental rate ($1/d$) and fecundity (F) to prey density, assuming that the rate of ingestion I is approximately proportional to the number of prey eaten, i.e.

$$I = k N_a \qquad (7.9)$$

where k is the proportion determined by the biomass of prey and the fraction of each item of prey that is utilized, can be derived using equations 7.4, 7.6 and 7.7:

$$F = \lambda [kaN/(1 + aT_hN) - c] \qquad (7.10)$$

$$1/d = \alpha [kaN/(1 + aT_hN) - \beta] \qquad (7.11)$$

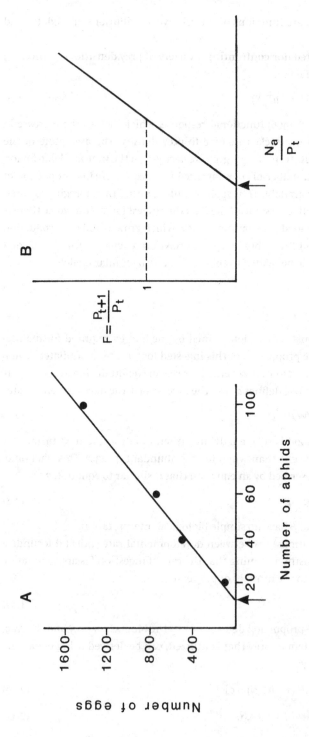

Fig. 7.5. The relationships between (A) adult lifetime fecundity of *Propylea japonica* and the number of aphids supplied per unit time (after Kawauchi, 1981) and (B) the predicted relationship between fecundity (F) and number of prey per predator (N_a/P_t). (Arrow indicates the minimum prey abundance for oviposition.)

In the case of the fecundity relationship (7.10) the empirical data is not fully supportive as the relationship for *Propylea japonica* (Kawauchi, 1981) is linear (Fig. 7.5) rather than curvilinear. However, this may be a consequence of the ladybirds not being satiated at the highest feeding levels in these studies. The study on *Adalia decempunctata* by Dixon (1959) indicates that the relationship is curvilinear for this species. The empirical data on the relationship between developmental rate and prey density is as equation (7.11) predicts, curvilinear (Fig. 3.1; p. 37).

Assuming mortality is solely attributable to starvation and that the frequency with which members of a population of predators die is normally distributed about a mean ingestion rate μ_1, with standard deviation σ_1 the proportion surviving, s, to complete development within any particular instar at an ingestion rate I is

$$s = \tfrac{1}{2}\,\pi \int_{-\infty}^{z} \exp\left(-z^2/2\right) dz \tag{7.12}$$

and

$$z = (I - \mu_1)/\sigma_1 \tag{7.13}$$

Using equations 7.4 and 7.9, survival (s) can be expressed in terms of prey density N (Lawton *et al.*, 1975). The empirical data available for ladybirds supports the prediction that the proportion surviving shows a curvilinear increase with prey density. Thus it would appear we have some understanding of the underlying biological mechanisms.

In contrast to parasitoids, which have a single stage that searches for hosts, the adult female, most predators also seek prey as juveniles. Thus search rates and handling times of predators vary with the stage of their development and size of prey. This complexity poses considerable difficulties when modelling predator–prey interactions. Therefore, it is important to ask whether increasing the complexity of the models to include all the developmental stages is useful or necessary. The practical objective of modelling insect predator–prey interactions has been to understand what determines the abundance of prey insects, particularly pest species. The elegant theoretical studies of parasitoid–host interactions have mainly contributed to our understanding of the factors that stabilize an otherwise unstable interaction. Notwithstanding the enormous intellectual effort and time that has been devoted to developing the theory of population dynamics, the practice of biological control is still more of an art than a science. For example, there is still no consensus about what constitutes a potentially good biological control agent (p. 197; Luck, 1990).

As the attempts to make the Nicholson & Bailey model more appropriate to predators have resulted in dauntingly complex mathematical models, which have contributed nothing to the practice of biological control, it may be time to consider a change of emphasis.

Optimal foraging approach

The interaction between ladybirds and aphids is well studied and the following model of Kindlmann & Dixon (1993) is based on the current understanding of this interaction, with the expectation that it will apply to other predator–prey interactions.

Aphid populations generally grow exponentially in the initial phases of development and on reaching a certain critical population density the aphids switch to alate production and/or migration. The switch to migration can be seen as a response to a deterioration in the quality of the host plant or severe intraspecific competition. The switch results in a rapid decline in colony size (Fig. 7.6(a)). Ladybirds arriving in an aphid colony are faced with an optimization problem, the maximization of offspring production. As indicated, an aphid colony is a temporary and finite food resource. In addition, other insect predators may lay eggs in the colony. It is assumed that ladybird larvae show a Holling type II functional response to aphid abundance and that the larvae will eat their conspecifics when aphids become rare. Because each aphid colony only exists for a finite period of time the larvae may complete their development but not reproduce in that colony.

The assumptions made by Kindlmann & Dixon (1993) in developing their model are:

(A1) Aphid populations grow exponentially with a growth rate r, which is the difference between the birth rate, b, and the death rate, d. The aphid population biomass is $x_b(t) = x_s x(t)$, in which average aphid size, x_s, is assumed to be constant, whereas the density, x, changes in time.

(A2) Density dependence comes into effect after a critical density is reached. When the aphid population exceeds the critical density, x_{crit}, the individuals switch to alate production and/or migration at the same rate, b. This switch, however, is not instantaneous and is described by the switch function (Fig. 7.6(b)).

(A3) The total biomass per unit area of ladybird larvae at time t is $y_b(t) = y(t) \cdot y_s(t)$. Thus, both their size, y_s, and their density, y, may change in time. However, the changes in density, y, are due to cannibalism, as they do not reproduce.

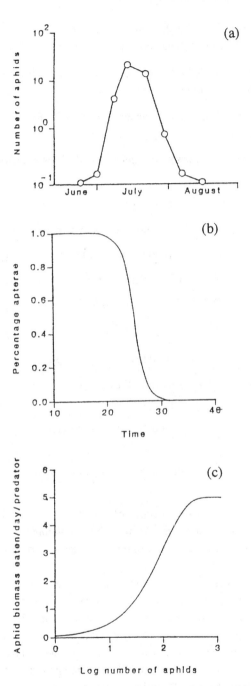

Fig. 7.6.(a) Empirical data on the trend in abundance of an aphid colony (after Kareiva, 1986); (b) function g (equation 7.18), describing the switch from aptera to alate production; (c) functional response of ladybirds to aphid density (equation 7.19).

(A4) The functional response shown by ladybird larvae (biomass eaten per predator per unit time) is of Holling type II (Fig. 7.6(c)).

(A5) The larvae attack aphids and conspecifics at random, but prefer aphids with the preference factor p.

(A6) The biomass eaten by the larvae is converted into their biomass with a conversion efficiency of q.

Assumptions (A1)–(A6) lead to a set of algebraic and differential equations:

	Describes	
$x^1 = (bg - d)x - p_x yf$	aphid population dynamics	(7.14)
$y^1 = -p_y yf$	number of ladybird larvae	(7.15)
$y^1 s = qf$	size of ladybird larvae	(7.16)
$x_b = xx_s$	biomass of aphids	(7.17a)
$y_b = yy_s$	biomass of ladybirds	(7.17b)
$g(t) = 1 - \dfrac{e^{vt}}{e^{vT} + e^{vt}}$	switch to alate production	(7.18)
$f(x,y,y_s) = m \cdot y_s[1 - e^{-a(x+y)}]$	functional response of ladybird larvae	(7.19)
$p_x = \dfrac{x}{x + y/p}$	proportion of aphids in diet	(7.20a)
$p_y = \dfrac{y}{p(x + y/p)}$	proportion of ladybirds in diet	(7.20b)

Simulations were performed using a range of parameter values, which either included all the possible values (q), most of the possible values (p, y_0, $y_s(0)$, a, m) or were varied by the choice of units (t, x_s, x_0) or only realistic values were chosen (b, d, T, v).

All the simulations resulted in a common prediction: there is an optimum number of ladybird larvae, which maximizes the biomass production of ladybird larvae at the end of the existence of an aphid colony. This is illustrated in Fig. 7.7. The sharp increase in ladybird biomass is followed by a sharp decline, when the optimum initial number of larvae is exceeded. This indicates that there should be a strong selection pressure for ladybirds to optimize the initial number of larvae.

In addition, if they do optimize the initial number of larvae then the simulations also indicate there is only a slight reduction in the peak

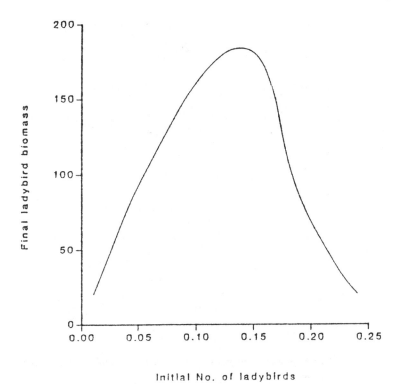

Fig. 7.7. Ladybird biomass production in relation to the initial number of ladybirds, as predicted by the model.

number of aphids (Fig. 7.8). That is, if ladybirds behave optimally they should not substantially reduce the size of aphid colonies.

This model does not explore the effect of increasing the size of the aphid colony on the optimum number of ladybird larvae and the percentage decrease in the peak number of aphids. In a laboratory study using larvae of *Coleomegilla maculata* and the aphid *Rhopalosiphum maidis* on sorghum Lorenzetti (personal communication) has both tested and extended the above model. Interestingly, increasing the number of aphids available did not result in a significant change in the optimum number of ladybird larvae or in their effect on the percentage decrease in the peak number of aphids.

The model can also be extended to a multiple patch system using the foraging theory patch model of Stephens & Krebs (1986). In this case the number of eggs laid in each patch of prey will depend on the relative abundance of adult predators and prey (unpublished results). When ladybirds are abundant relative to their prey the initial number of eggs per

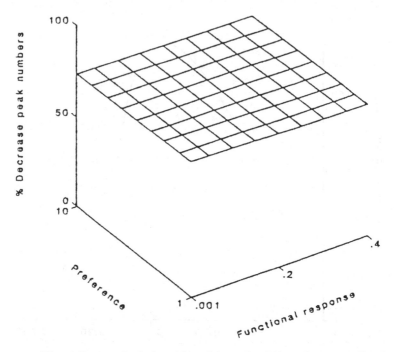

Fig. 7.8. Decrease in the logarithm of the peak aphid numbers, as predicted by the model.

patch is likely to be higher than when ladybirds are scarce. Cannibalism is then likely to result in the death of proportionally more of the eggs and larvae.

The developmental time of a ladybird is considerably longer than that of an aphid and comparable in length to the average time for which an aphid colony exists. Thus it does not make sense for ladybirds to oviposit in old aphid colonies as their larvae will not be able to complete their development. Therefore, there is also a great selective advantage in adults being able to detect aphid colonies in the early stages of development.

This poses the question: is it reasonable to expect ladybirds to have this level of omniscience? The empirical data indicate that they do lay their eggs early in the development of an aphid colony and cease laying when the risk of their eggs being eaten by larvae is high. They do this by responding to simple cues (p. 106). When ladybirds are abundant then they are more likely to lay more than the optimum number of eggs being laid in many patches. Subsequent egg and larval cannibalism assures the survival of at least some of the larvae. Alternatively, laying more than the

optimum number of eggs may be an effective way of harvesting a transi-
tory resource or the best strategy in view of the uncertainty about the
future quality of the patch (p. 147).

Minimalism – generation time ratios

The increasing complexity of the classical approach has resulted in
rhetorical rather than empirical problems dominating the field of population dynamics. A refocusing on the empirical problem might be
achieved by a process of minimalism. Following Slobodkin (1986), this is
the process of choosing to work in the simplest possible mode that is still
recognizable in terms of and relevant to the assumptions that are made
about the processes that drive predator–prey dynamics.

The death rate imposed by a predator population on a prey population is linked to the rate at which predators locate and consume prey, and
hence to the rate of increase of the predator population. That is, the
values of the various life history parameters of a predator are important
in determining the outcome of predator–prey interactions. Figure 7.4
also clearly indicates that the equilibrium density of the prey is determined not only by the attack rate of the predator but also by the rate of
increase of the prey. In prey organisms like aphids it is clear that their rate
of increase is positively correlated with their growth rate, which is positively correlated with their developmental rate. That is, the life history
parameters of interest to population ecologists are all correlated with one
another. This is fundamental to the minimalistic approach developed
below. It is likely that this is true for all organisms. Similarly, other features of potentially successful biocontrol agents such as a high search
rate and a marked ability to aggregate in patches of high prey density
(Hassell, 1978) are also likely to be positively correlated with other life
history parameters. This leads to the concept that some species generally
develop, grow, reproduce and search very rapidly and others slowly (p. 80).
This is often associated with size but as indicated in Chapter 4 this is not
always the case.

As the equilibrium density of the prey is determined by both the
attack rate of the predator and the rate of increase of the prey, and these
features of predators and prey are positively associated with other life
history parameters of these organisms it should be possible to produce a
simpler model of predator–prey interactions. Such an approach was
adopted by Janssen & Sabelis (1992) for mites. They show that life history
parameters of mites are closely correlated with one another and argue
that the pest status of herbivorous mites and their control by predatory

mites are dependent on their relative population growth rates (r_m). They also indicated that although rates of population increase are usually determined under optimum conditions, which would appear to make their value under field conditions debatable, nevertheless a predatory mite that has an r_m equal to or greater than its prey should at least have the potential of reducing the abundance of its prey. However, they regard r_m and predation (attack) rate as operating independently of one another and subject to different selection pressures. That is, r_m may be compensated for by a higher attack rate. This is a widely held view, which is difficult to rationalize and lacks empirical support (p. 80).

There are further arguments for simplifying models of predator-prey interactions. If adult predators are successful in assessing patch quality then it is likely that their larvae experience feeding conditions equivalent to those experienced on the asymptotic part of the functional response curve. Thus, initially larvae possibly experience relatively little variability in the availability or quality of food , which therefore may not be as important in determining their developmental and survival rates as other factors such as temperature. Mills (1982b) noted that 'despite the obvious importance of temperature, little interest has been shown in the influence of temperature on predator-prey interactions'. Laboratory results indicate that temperature has a more marked effect on the range in larval development rates within a species than availability of food (Fig. 7.9) with the exponent on temperature for four species ranging from 2.0 to 3.3 and on food supply for three species from 0.94 to 0.96. That is, they show a much greater range in developmental rates in response to changes in temperatures than to changes in food supply. Similarly, changes in temperature affect the developmental rate of aphids more markedly than do changes in food supply (Dixon, 1987). That is, the developmental rates of both predator and prey are very markedly affected by changes in temperature, a feature that has been ignored in all theoretical treatments of predator-prey interactions.

Accepting that the r_m of insect predators and prey are markedly affected by temperature then there is a strong argument to follow Janssen & Sabelis (1992) and simplify the conceptual basis of the predator-prey interaction. One such approach is to look for a life history parameter that is easy to measure and correlated with all the other life history features of an organism over a wide range of temperatures. The developmental rate appears to be ideal for this purpose, as there are very strong relationships between r_m, and temperature, and r_m and developmental rate within species (Stäubli Dreyer et al., 1997b) and between species of ladybirds (Fig. 7.10), and developmental rate is very easily measured.

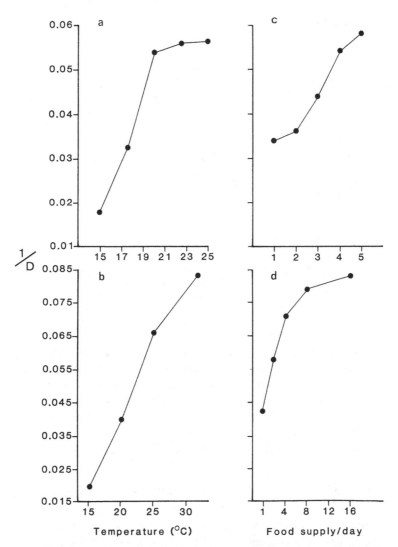

Fig. 7.9. Developmental rate in relation to temperature and food supply for *Adalia bipuncta* (a and c) and *Harmonia axyridis* (b and d). (Data from: Mills, 1979; Schanderl *et al.*, 1985.)

Studies on aphid–ladybird systems indicate that the prey has a much shorter developmental time than the predator (Dixon *et al.*, 1997) and that the abundance of prey in each patch varies greatly in time even in the absence of predators (cf. Fig 7.6(a); Dixon, 1997). The optimum oviposition strategy of a predator in such a system is likely to be determined by expectations of future bottlenecks in prey abundance. The strategy of a

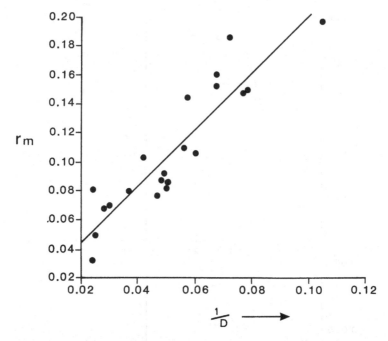

Fig. 7.10. Intrinsic rate of natural increase (r_m) in relation to developmental rate ($1/D$) for various species of ladybirds. (Data from: Chazeau, 1981; Chazeau *et al.*, 1991; Ding-Xin & Zhong-Wen, 1987; El Hag & Zaitoon, 1996; Fabres & Kiyindou, 1985; Gibson *et al.*, 1992; Gutierrez & Chazeau, 1972; Kairo & Murphy, 1995; Kanika-Kiamfu *et al.*, 1992; Kawauchi, 1983, 1985; Napompeth & Maneeretana, 1990; Obrycki *et al.*, 1993; Stäubli Dreyer *et al.*, 1997b; van Steenis, 1992; Wright & Laing, 1978.)

predator with a long larval developmental time will depend on a longer projection of the future prey abundance in a patch and will therefore include more bottlenecks compared to a predator with a short developmental time. That is, from a slow-developing predator's perspective the prey is fluctuating dramatically in abundance, i.e. the prey dynamics appear jagged.

The dynamics of such predator–prey interactions have been explored by means of a model (Kindlmann & Dixon, 1999a) where the predator-free prey density at time t, x_t, is assumed to be a random walk:

$$x_{t+1} = x_t + \epsilon \tag{7.22}$$

where ϵ is a random number between $-e$ and e. This includes both the intrinsic dynamics of the prey and the influence of environmental conditions, of other species, etc., and bears some resemblance to field data

(Kindlmann & Dixon, 1999b). The prey generation time is therefore 1 time unit and the parameter e is a measure of the degree of jaggedness in the prey dynamics. The primary objective of this was, however, to create jagged prey dynamics, with some, rather than an absolute, resemblance to reality. The generation time of the predator is D time units and it needs v prey items for successful completion of its larval development. The predator numbers only change when $t \bmod D = 0$, i.e., when the larvae become adult, leave the patch and different F females arrive to lay their eggs. The number of eggs laid by one female per patch is assumed to be equal to:

$$y_t = \frac{m}{v(F+1)} \tag{7.23}$$

where $m = \min_{\tau \varepsilon <t, t+D>} x_\tau$, a consequence of which is that the size (fitness) of the larvae is positively correlated with the minimum expected number of prey in the absence of predators during larval development, relative to their voracity.

The simulations for different generation time ratios in Fig. 7.11 clearly illustrate that the effect on the prey population density is inversely related to the relative developmental time of the predator. That is, the outcome is determined by the relative speeds of the predator and prey dynamics, and none of the specific assumptions (discrete time, predator-free prey dynamics, parameter values) affect the outcome. This indicates that an effective biological control agent must develop faster or at a rate similar to that of its prey or host.

In summary, the application of foraging theory to predator–prey dynamics has focused attention on the adaptive significance of the foraging and oviposition behaviour of predators. Experimental studies have identified mechanisms that enable predators to forage in ways that often come close to optimal. Above all this approach revealed the importance of the generation time of the predator relative to that of its prey. Simplification of predator–prey dynamics to just the relative generation times of the predator and prey could be useful in resolving another important problem: what are the attributes of an effective biological control agent? This approach indicates that the most important attribute is likely to be its rate of development relative to that of its prey. The value of this approach is that it reveals it is the relative rates of development of the predator and prey, not their absolute values, that are important. If the developmental rate of the predator is similar or faster than that of the prey then the predator is potentially capable of suppressing the prey. This is referred to as the generation time ratio.

In the development of models of parasitoid–host interactions little

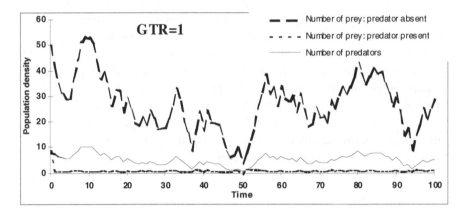

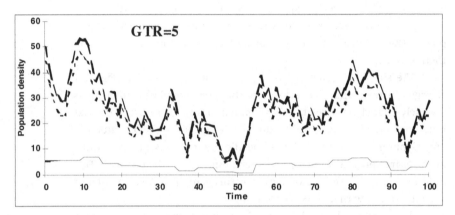

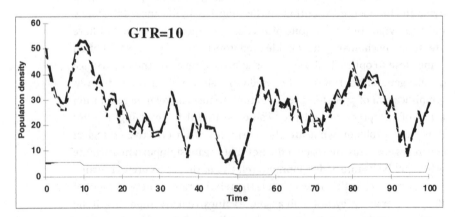

Fig. 7.11. The effect of different generation time ratios (GTR) on the abundance in time of the prey, in the absence and presence of predators, and of the predator. (Parameters: $K = 10$, $v = 5$; for $t \bmod D = 0$; $F_t = y_{t-1}$; for $D = $ GTR). (After Kindlmann & Dixon, 1999a.)

attention has been paid to the relative developmental times of host and parasitoid, possibly because in most cases they have similar developmental times. However, there is one study that is directly relevant to the idea presented here. Godfray & Hassell (1987) use a model to show that the relative lengths of host and parasitoid generations have a profound affect on population dynamics. Their study concentrates on values of generation time ratio less than 1.0 and shows that with shorter parasitoid than host generation times, host population size and variability increases. The potentially important implication for biological control is appreciated, especially the pragmatic point that developmental time is one of the easiest attributes to measure under natural conditions. However, this study does not appear to have been followed up or its significance appreciated by practitioners of biological control. This may be because there is a tendency to view each life history parameter as independently subject to selection. For example, Janssen & Sabelis (1992) suggest that predatory mites with a lower r_m than their prey can nevertheless still effectively control the abundance of their prey if they have a high voracity. The implication of this is that voracity has evolved independently of the other attributes, like developmental time, that are important in determining r_m. The big question is: to what extent can life history parameters be viewed as independent of one another? Food consumption, relative to growth rate and developmental rate, would appear to be interdependent. Assuming that food supply is limiting there is likely to be strong selection leading to the maximization of assimilation rate, which should result in a strong link between rate of food consumption and rate of growth. In addition the data on ladybirds tend to argue in favour of life history parameters being linked (p. 80), i.e. we are dealing more with a suite of parameters than with independent parameters. If this is the case then focusing on one easily measured parameter, like development time, can be justified. It is also a very simple pragmatic solution to modelling predator–prey interactions.

In conclusion, ladybirds are classical insect predators, as their larvae pursue and subdue their prey. In contrast to parasitoids, which have a single stage that searches for hosts, the adult female, most predators seek prey as juveniles. Their search rates and handling times vary with stage of their development and size of prey. This complexity poses considerable difficulties when modelling predator–prey interactions. An optimal foraging approach to ladybird–prey interactions has revealed the importance of the relative developmental rates of predator and prey.

There is strong evidence that all the life history parameters of interest to population ecologists are interdependent. Thus it is reasonable as well as practical to focus on one easily measured parameter, like developmental time. This has resulted in a simplification of the conceptual basis of predator–prey interactions, which indicates that if the developmental rate of a predator is similar or faster than that of its prey then the predator is potentially capable of dramatically reducing the abundance of its prey.

8

Intraguild predation

An abundance of prey generally attracts large numbers of a diverse array of natural enemies. For example, two or more species of anthocorid bug, hoverflies, ladybirds, mirid bugs and parasitoids may all attack a population of sycamore aphids, *Drepanosiphum platanoidis*, at one and the same time (Fig. 8.1). As regardless of differences in taxonomy or tactics they all exploit the prey in a similar way then following Root (1967) they are seen as constituting a guild. In jointly exploiting a patch of prey the members of a predatory guild may affect each others' foraging success. The interactions, however, differ from competition in that one participant (the predator) accrues immediate energetic gains, and from classical predation in that it reduces potential competition. Intraguild predation occurs when one species feeds on another in a predatory guild. The aggressor is the intraguild predator, the victim the intraguild prey, and the common resource is the extraguild prey.

The ecology and evolution of intraguild predation has been extensively reviewed by Polis *et al.* (1989) and Polis & Holt (1992). In terms of ladybirds the best studied and understood are the aphidophagous guilds. Because of the way these guilds are structured (p. 106) the most important interactions are likely to occur between the immatures of the different species of predator.

GUILD STRUCTURE

A major factor limiting our understanding of the evolution of intraguild predation is the complete lack or poor knowledge of the preferred habitats of even our commonest ladybirds. Few authors have addressed this problem; however, a notable exception is Honěk (1985). His studies

173

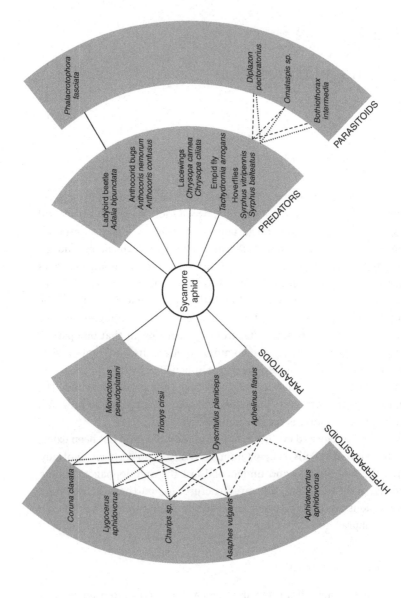

Fig. 8.1. The predators, parasitoids and hyperparasitoids of the sycamore aphid, *Drepanosiphum platanoidis*.

have revealed that there are overlaps in habitat preferences, i.e. some species are more likely to interact with one another than with other species (Table 5.1, p. 98). Another approach is to determine the relative abundance of the different coccinellid species and other aphidophaga in particular habitats. Accepting that the results obtained are typical for the habitat concerned then each would appear to be dominated by one particular species of predator (Fig. 8.2). If this is shown to be generally true then within any particular habitat the biggest threat to the commonest species of predator is cannibalism (p. 134). The relatively less abundant species in the guild, however, are mainly at risk of being attacked or eaten by the more abundant species. That is, under natural conditions one would expect the threat from cannibalism to be more important for the common and intraguild predation for the less common species. In addition, there can be big differences in size between the smallest and largest species of predator. As the victim is usually the smaller individual one would expect the smaller species to invest more in defence. That is, the relative abundance and size of the species of ladybird in each habitat is likely to have shaped the evolution of defences in these predators. In addition, when the extraguild prey (aphids) become scarce the incidences of both cannibalism (p. 141) and intraguild predation are likely to increase. For example, in years when the sycamore aphid is uncommon in summer the parasitized aphids are then an important source of food for the nymphs of two predatory bugs, *Anthocoris confusus* and *A. nemorum* (Dixon & Russel, 1972).

PREDATOR–PREDATOR INTERACTIONS

Whether to use natural enemy complexes rather than a single natural enemy in biological control programmes is a long-standing issue (Turnbull & Chant, 1961; Hassell & Varley, 1969: Ehler, 1990; Benrey & Lamp, 1994). There are cases in which enemy complexes are thought to provide enhanced pest suppression (Frazer *et al.*, 1981; Murdoch, 1990), but there are also instances in which predator complexes appear to be less effective than single natural enemies in reducing pest populations (Rosenheim *et al.*, 1995).

Additive/non-additive effects

Researchers interested in the use of predators as biocontrol agents have done experiments to determine whether they act in a simple additive manner or whether the addition of another species of predator

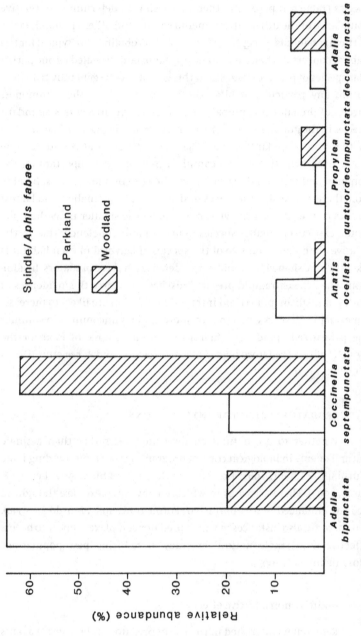

Fig. 8.2. Relative abundance of five species of ladybirds feeding on *Aphis fabae* on spindle in parkland and woodland. (After Barczak *et al.*, 1996.)

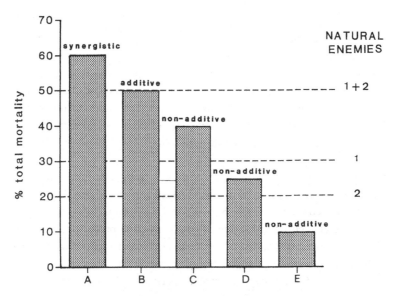

Fig. 8.3. The potential range of percentage total mortality inflicted by two
(1 + 2) natural enemies on their host population. It can be more than the
sum (A – synergistic), equal to the sum (B – additive), or less than the sum to
varying degrees (C, D and E – non-additive) of the mortalities inflicted by 1
and 2 acting on their own. (After Ferguson & Stiling, 1996.)

enhances or reduces the ability of already existing predators to reduce
the numbers of the pest. That is, it is of considerable practical and theoretical importance to know whether the total mortality inflicted by a
guild of natural enemies on a prey population is greater than the sum of
the mortalities inflicted by each of the natural enemies, i.e. is there a synergistic effect, or is the combined effect additive, or is it less than the sum
of the individual mortalities, i.e. non-additive or antagonistic to varying
degrees (Fig. 8.3). Simple laboratory experiments in which the larvae of
two species of predator are placed in Petri dishes, with and without
aphids, clearly indicate that generally the larger predator is likely to eat
the smaller irrespective of species and that the presence of aphids (extraguild prey) reduces the likelihood of intraguild predation (Sengonca &
Frings, 1985; Lucas *et al.*, 1997a, 1998; Phoofolo & Obrycki, 1998). Less artificial experiments in which the larvae of two species of predator are
placed on a plant with and without aphid prey have also been done. One
such study on three species of larval aphidophaga; a fly, *Aphidoletes aphidi-
myza*, a lacewing, *Chrysoperla rufilabris* and a ladybird, *Coleomegilla maculata lengi*, confirmed the occurrence of intraguild predation and that it
tends to be asymmetrical with the larger individual the intraguild

predator, and that the incidence of intraguild predation decreases when extraguild prey are present (Lucas *et al.*, 1998). Other studies were set up not to test for intraguild predation but to determine whether equal numbers of two predators depress prey population density more or less than twice as many predators of either species alone. Those of both Evans (1991) and Chang (1996) are consistent in not revealing an interaction between the predator species. In the case of Chang's experiments this may have been due to the spatial separation on the plants of the two species of predator. Under the conditions of moderate competition for prey prevailing in Evans's experiment, the weight gains of initially similar-sized predators over a period of 45 hours did not depend on whether the competitors were heterospecific or conspecific. That is, these less artificial experiments did not reveal intraguild predation. However, although undoubtedly more realistic than the experiments done in Petri dishes, nevertheless the ladybirds and chrysopids used in these experiments were similar in size and did not experience a shortage of prey, both of which are likely to affect the outcome under more natural conditions.

Most field and cage studies indicate that the total mortality inflicted by an aphidophagous guild is non-additive and that pest densities are often lower when attacked by a few than by many natural enemies. Some ladybirds, like *Rodolia*, are reluctant to eat their prey when it contains the immature stages of a parasitoid (Quezada & DeBach, 1974). However, most readily eat parasitized prey. For example, *Hippodamia convergens* destroys 73–100% of the immature stages of the parasitoid *Lysiphlebus testaceipes* developing in populations of the cotton aphid, *Aphis gossypii* (Colfer & Rosenheim, 1995). Similar, non-additive mortality has been recorded for aphid populations attacked by *Coccinella septempunctata* and the parasitoid *Aphidius ervi* (Taylor *et al.*, 1998), *Cycloneda sanguinea* and *A. floridaensis* (Ferguson & Stiling, 1996) and between ladybirds and the parasitoids of mealybugs and scale insects (Prakasan & Bhat, 1985; Gutierrez *et al.*, 1988; Samways & Wilson, 1988). This does not always result from ladybirds eating the immature stages of parasitoids. It can result from ladybirds reducing the effectiveness of ovipositing parasitoids, which is usually referred to as interference, or to interaction modification, in which the previous presence of a ladybird can cause parasitoids to leave a patch (Ferguson & Stiling, 1996 ; Taylor *et al.*, 1998). If the oviposition-deterring pheromones produced by many predators prove to be oviposition-deterring *alomones* as claimed by Růžička (1996, 1997*a*, *b*) and Růžička & Havelka (1998) then the tendency of aphidophaga to avoid patches of prey already under attack by natural enemies is likely to be of common occurrence. Therefore, models in which the individual species' potential consumption of prey are simply added (e.g. Freier *et al.*, 1998) are likely to greatly overestimate the role of natural enemies in reducing the abundance of pests.

Predator facilitation

The above studies indicate that if two predators do not interact then their combined impact on the prey population will be additive, i.e. equal to the sum of their individual impacts. If one predator species kills or affects the other predator's foraging behaviour, then the combined impact is non-additive. However, predator species can also interact synergistically – a phenomenon termed 'predator facilitation' (Charnov *et al.*, 1976). This can occur when the foraging activity of one predator species alters the behaviour or feeding niche of the prey, making it more susceptible to attack by other predator species.

Many species of aphid respond to the presence of a foliar-foraging predator by dropping to the ground where they are at risk from ground predators. Thus such a system would appear to be ideal for studying predator facilitation. In a series of laboratory and field experiments Losey & Denno (1998) investigated the interactions between the foliar-foraging *Coccinella septempunctata* and ground-foraging *Harpalus pennsylvanicus*, predators of the pea aphid. The combined impact of these predators was nearly double the sum of their individual predation rates (Fig. 8.4). The mechanism for the interaction was the aphid 'dropping' behaviour elicited by *C. septempunctata*, which rendered the aphid susceptible to predation by *H. pennsylvanicus* on the ground. Although the result is clear it is worth asking to what extent is it a consequence of the experimental design. For example, how realistic are the predator/prey ratios used in these experiments? If by virtue of dropping to the ground in the presence of foliar-foraging predators an aphid exposes itself to a much higher risk of predation then such behaviour would appear to be maladaptive. Thus, although it is interesting one needs to be convinced of the relevance of this result. Hopefully this study will stimulate more realistic field experiments.

Top predators

The action of a top predator in regulating the abundance of a natural enemy, which previously controlled the prey population, can result in what is referred to as a trophic cascade. For example, it is thought that the generalist nabid predator *Zeleus renardii* reduces the

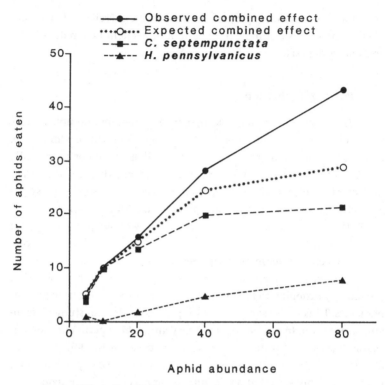

Fig. 8.4. Effect of aphid abundance and predator treatment on the number of aphids eaten. (After Losey & Denno, 1998.)

effectiveness of the chrysopid *Chrysoperla carnea* in regulating the abundance of the cotton aphid, *A. gossypii*. This results in aphid outbreaks, which adversely affect the performance of the host plant, cotton (Rosenheim *et al.*, 1993; Cisneros & Rosenheim, 1997; Rosenheim *et al.*, 1999). That is, the top predator frees the prey from regulation by the primary predator allowing the herbivore to increase in abundance and regulate the abundance of its host plant.

Although many ladybirds are polyphagous they do tend to specialize on mainly either aphids or coccids. Some even are specific predators of mycophagous ladybirds. Thus within guilds of ladybirds there could be species that as well as depending on extraguild prey have also specialized in capturing and eating other species of the guild. This has been tested using three species of ladybird, *Coccinella septempunctata brucki*, *Harmonia axyridis* and *Propylea japonica*. They were either reared from the second instar on their own or with equal numbers of the other two species on small shrubs infested with the same numbers of aphids. In each case the

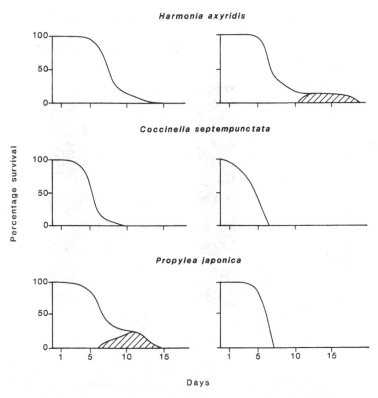

SINGLE SPECIES   MIXED SPECIES

Harmonia axyridis

Coccinella septempunctata

Percentage survival

Propylea japonica

Days

Fig. 8.5. The percentage survival of Harmonia axyridis, Coccinella septempunctata and Propylea japonica when reared from the second instar on small shrubs of Hibiscus syriacus each equally infested with Aphis gossypii. Nine larvae of each species of ladybird were reared as a group on a plant (single species), or in groups consisting of three larvae of each of the species (mixed species). The shaded area indicates that some of the larvae survived to the pupal stage; the aphid population became extinct on about day 5 in each case. (After Sato, unpublished.)

total number of ladybirds was the same. When competing with the other species both *C. s. brucki* and *P. japonica* did worse than when on their own, especially *P. japonica*, which when on its own was the only species that survived to the pupal stage. Most surprising was that *H. axyridis* survived to the pupal stage when competing but not when reared on its own. This indicates it is effectively using the other species as a source of food (Fig. 8.5; Sato, unpublished data). That is, the results of this series of experiments are supportive of the notion that very strong interactions will

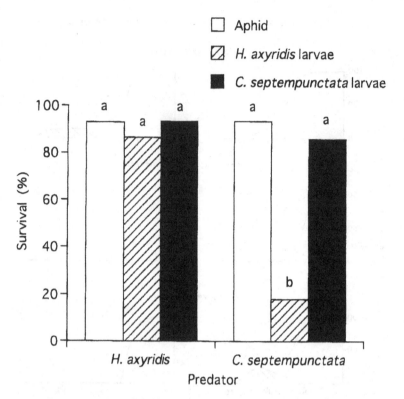

Fig. 8.6. Survival of the fourth instar larvae of two species of ladybird fed aphids, *Harmonia axyridis* larvae or *Coccinella septempunctata* larvae. (Bars labelled with the same letter are not significantly different). (After Yasuda & Ohnuma, 1999.)

occur between different species of ladybirds attacking the same resource, especially as when towards the end of their development aphids are likely to become a limiting resource. In addition, it would appear that *H. axyridis* is a top predator regularly attacking and eating other species of ladybirds. The interactions between species may more often result in one of the species leaving the patch or plant then being killed. That is, the reaction results in the displacement of a potential competitor.

If a top predator then *H. axyridis* should be well adapted to survive and thrive on diets of both aphids and other species of ladybirds. The study of Yasuda & Ohnuma (2000) lends support to this notion, as fourth instar larvae of *H. axyridis* fed on diets of aphids and ladybird larvae survived and grew equally well, whereas those of *C. septempunctata* did better on a diet of aphids than on a diet of other species of ladybird larvae (Fig. 8.6). If a top predator one would also expect *H. axyridis* to prefer to oviposit in patches of prey that are already being attacked by other species of ladybirds and that its eggs and larvae should be well defended against intraguild predators. In alfalfa fields *H. axyridis* arrives after the other ladybirds and eats the prepupae and pupae of *C. septempunctata brucki* (Takahashi, 1989), and the spiny covering and large mandibles would appear to protect the larvae of *H. axyridis* against intraguild predators. More studies on the predatory relationships between *H. axyridis* and other species of ladybird with which it usually co-occurs in the field are needed to confirm the trophic position of this predator. If confirmed the intro-duction of this species into other countries for use in biological control could seriously affect the abundance of native ladybirds, although there is no evidence of this (cf. p. 196).

COST OF INTRAGUILD PREDATION

Intraguild predation is mainly considered in terms of the advan-tages to the predator in increasing the availability of food and reducing competition. However, intraguild predation could have costs other than the risk to the predator of becoming the prey. The rich array of alkaloids present in all stages, from egg to adult, have mainly been seen as a defence against visual-hunting vertebrate predators. The fact that the adults of many aphidophagous species appear to be aposematically coloured has fuelled this view. However, many if not all the species that are not aposematically coloured also contain alkaloids. Thus it is likely that these chemicals could be important in protecting ladybirds against invertebrate predators.

Generally the eggs of animals are seen as a rich source of food in contrast to the seeds of many plants, which often contain highly toxic compounds. The first study to challenge this concept was that of Agarwala & Dixon (1992). They offered starving adults and larvae of two species of *Adalia* (*A. bipunctata* and *A. decempunctata*) and two species of *Coccinella* (*C. septempunctata* and *C. undecimpunctata*) their own and the other species' eggs. The *Adalia* species consumed the eggs of the other species of *Adalia* and those of the two species of *Coccinella* equally readily, whereas the *Coccinella* species were more reluctant to eat the eggs of the *Adalia* species. This effect is due mainly to the great reluctance of *C. sep-tempunctata* to eat the eggs of *A. bipunctata*. In terms of the proportion of the eggs eaten by the other species *A. bipunctata* is the least and *C. undecim-punctata* the most readily eaten (Fig. 8.7). Similarly, large larvae of the *Coccinella* species are more reluctant to eat small larvae of *Adalia* species than vice versa.

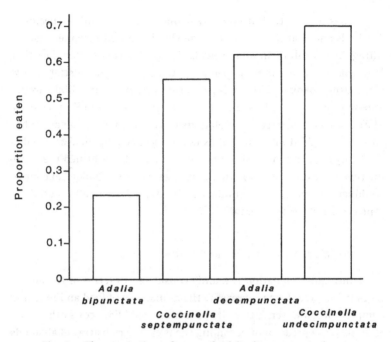

Fig. 8.7. The proportion of eggs of *Adalia bipunctata, A. decempunctata, Coccinella septempunctata* and *C. undecimpunctata* eaten by the other species of this group of ladybirds. (After Agarwala & Dixon, 1992.)

This reluctance to eat the eggs and larvae of other species is adaptive because although their own eggs are a high-quality food those of other species may be toxic. For example, larvae of *C. septempunctata* die if fed only eggs of *A. bipunctata*. Similarly, few larvae of *A. bipunctata* fed only eggs of *C. septempunctata* survived and their development was prolonged compared to those fed conspecific eggs. After feeding on an egg of *C. septempunctata* larvae of *A. bipunctata* were frequently observed to vomit a black liquid (Hemptinne *et al.*, 2000b). As larvae and adults of these two ladybirds are similarly reluctant to eat their own eggs when painted with a water extract of crushed eggs of other species, it is likely they are protected by the alkaloids they contain (Agarwala & Dixon, 1992). The adaline present in the eggs of *A. bipunctata* appears to be more effective against the larger *C. septempunctata* than is the coccinellin in its eggs against the smaller *A. bipunctata*. As the eggs and larvae of these two species frequently occur together in the field one would expect the smaller species to be better defended than the larger species. The novelty is that the defence is chemical.

Other studies on ladybirds that frequently co-occur in the field have

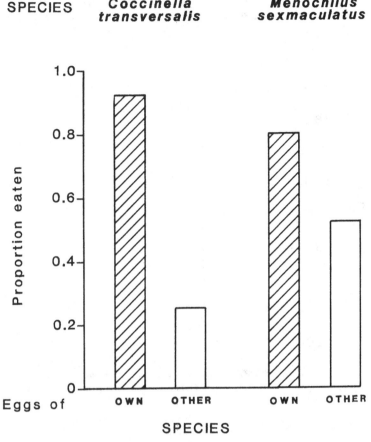

Fig. 8.8. The proportion of its own and the other species eggs eaten by *Coccinella transversalis* and *Menochilus sexmaculatus*. (After Agarwala *et al.*, 1998.)

revealed the same asymmetry in defense. The adults of *C. transversalis* show a much greater reluctance to eat the eggs of the smaller *Menochilus sexmaculatus* than the latter does of eating those of *C. transversalis*. This is revealed by both the proportion of eggs eaten and the time that elapses between a larva first encountering and eating the eggs (Figs. 8.8, 8.9; Agarwala *et al.*, 1998). Similarly, the larvae of *Calvia decemguttata* are more reluctant to eat the eggs of the smaller *C. quatuordecimguttata* than the reverse. In this case not only has the larger species a size advantage, its larvae have proportionally longer legs and are therefore able to move faster than the larvae of the smaller species (Vanhove, 1998; Hemptinne *et al.*, 2000b).

Hungry larvae readily attack and eat conspecific eggs but often less

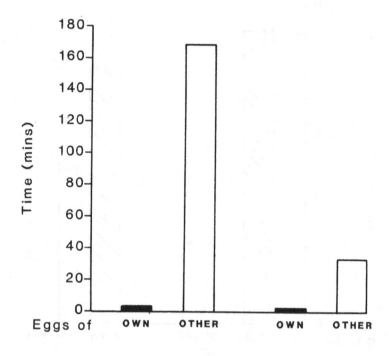

Fig. 8.9. The time that elapses between a larva first encountering and eating its own (■) species eggs and those of the other species (□) in *Coccinella transversalis* and *Menochilus sexmaculatus*. (After Agarwala *et al.*, 1998.)

readily those of other species of ladybird. However, a well-fed larva often appears to be very reluctant to attack and eat, even conspecific eggs. On encountering conspecific eggs such larvae immediately withdraw and move off in another direction. Interestingly, if the eggs are washed by immersing them in hexane for 2 minutes, which removes the waxy outer covering but does not otherwise damage the eggs, as most subsequently hatch, they are more readily attacked and eaten by well-fed larvae (Fig. 8.10). If the hexane extract of the surface of eggs is painted on to washed eggs larvae respond to these eggs as if they were unwashed (Fig. 8.11; Gauthier, 1996). Similarly, the waxy covering can be reciprocally transferred between the eggs of *A. bipunctata* and *C. septempunctata. A. bipunctata*

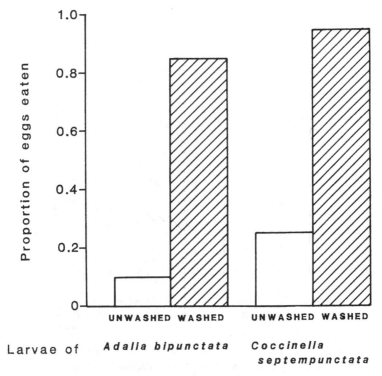

Fig. 8.10. The proportions of unwashed and washed conspecific eggs eaten by well-fed larvae of *Adalia bipunctata* and *Coccinella septempunctata*. (After Gauthier, 1996.)

larvae are more likely to attack and eat conspecific eggs than conspecific eggs that have been washed and painted with the surface extract from the eggs of *C. septempunctata*, and vice versa (Hemptinne *et al.*, 2000*b*). That is, larvae appear to be responding to species specific cues present in the waxy covering of the eggs (Table 8.1). Interestingly, the cue is qualitatively similar to the mixture of alkanes that deters adults from ovipositing in areas where larvae of their own species are already present (p. 107) and that which enables mating adults to recognize individuals of the same species (Hemptinne *et al.*, 1998). That is, for each species of ladybird a particular set of alkanes acts as a context specific signal.

In cannibalism and intraguild predation it is possible that the species-specific signals operate as follows. To attack an object covered in the species-specific group of alkanes could be hazardous for a well-fed larva because it could become the victim. If starving, however, then to attack and eat a conspecific could be advantageous because if successful the attacker will prolong its survival and reduce competition. If it

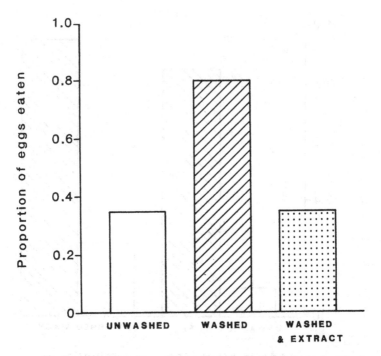

Fig. 8.11. The proportion of unwashed, washed, and washed and painted conspecific eggs eaten by well-fed larvae of *Adalia bipunctata*. Washed eggs were painted with an hexane extract of the surface waxes of conspecific eggs. (After Gauthier, 1996.)

encounters an object whose waxy covering is made up of an unfamiliar set of alkanes, then to attack could be extremely hazardous as not only could the attacker become the victim it could, if successful, eat something that is toxic. Therefore, if well fed it should avoid attacking and eating the immature stages of other species of predator not only because there is a risk of becoming a victim but also because of the possible adverse affect on its survival and rate of development. However, if starving then such an attack may prolong its survival sufficiently to enable the larva to locate more preferred food. That is, the species-specific spectrum of alkanes on the surface of eggs and larvae act as a cue; if familiar and you are hungry attack and eat if possible, if unfamiliar leave alone unless starving because it could be harmful.

In conclusion, guilds of aphidophaga appear to be dominated by one particular species. If general, then the greatest threat to the survival of the most abundant predator is likely to be cannibalism and to the other less abundant species of predator intraguild predation.

Table 8.1. *Concentration, in μg per 50 eggs, of the hydrocarbons in the n-hexane extract of the surface covering of the eggs of* Adalia bipunctata *and* Coccinella septempunctata

Hydrocarbons	A. bipunctata	C. septempunctata
C21	0.5	
C22	0.1	0.04
C23	0.3	0.13
7Me-C23	1.2	
9Me-C23	2.1	
C24		0.11
C25		0.17
C26		0.05
C27		0.44
C28		0.04
C29		0.12
Total	4.2	1.10

Although there are no rigorous field studies of the incidence and significance of intraguild predation in aphidophagous guilds the laboratory studies indicate that the combined effect of natural enemies is likely to be non-additive/antagonistic rather than additive or synergistic. There is also evidence that some ladybirds survive better if they are able to exploit not only aphids but also the immature stages of other aphidopahaga, i.e. they are top predators. Intraguild predation as well as bringing benefits in terms of an immediate energetic gain and reduced competition may also have a cost, other than the risk to the predator of becoming the prey. Ladybirds contain species-specific toxic alkaloids, which can adversely affect the fitness of other species of ladybird. The species-specific cuticular alkanes appear to signal these costs.

9

Biological control

The term 'biological control' was introduced by Smith (1919) to describe the use of natural enemies to control insect pests. It is defined as the regulation by natural enemies of another organism's population density at a lower average than would otherwise occur (DeBach, 1974). This definition does not indicate the degree of control or that it involves manipulation by man. The control of pests below the level at which they cause economic damage (economic threshold) by the deliberate introduction of exotic natural enemies is referred to as classical biological control. The designation 'classical' is in recognition of its relatively early first use in the 1800s. Although the origins of many pests are uncertain the impression inferred from the literature is that success is largely dependent on obtaining natural enemies of the pest from its area of origin. When it is carried out with due care, this is a lengthy and expensive process (Waage & Mills, 1992). Classical biological control, however, is very cost-effective because once established the system is self-sustaining.

Pests can also be controlled by the propagation and local release of large numbers of a natural enemy, which is often indigenous. Although this can be effective in controlling a local outbreak of a pest, especially if the stage of the natural enemy released is unable to leave the area or the outbreak is in a glasshouse, each outbreak has to be treated separately. Maintaining the facilities to produce large numbers of natural enemies for these inundative or augmentative releases, especially if the pest(s) does not exceed its economic threshold every year, is likely to be very expensive. In addition, the numbers of natural enemies can be increased locally by cropping practices that provide alternative prey or hosts for them to attack and propagate on before the target pest exceeds its economic threshold. The expectation is that these natural enemies will switch to attacking the pest and prevent it from becoming abundant enough to inflict economic damage. Both are included under the heading of augmentative biological control (Hoy, 1992).

The first statement of the rationale behind biological control is attributed to Erasmus Darwin, the father of Charles Darwin (Nicol, 1943). In a book entitled *Phytologia or the Philosophy of Agriculture and Gardening* published in 1800 Erasmus Darwin suggested that the eggs of hoverflies (syrphids) should be collected and kept over winter and the resultant larvae used the following year for controlling plagues of aphids. That is, he was proposing the augmentative release of an indigenous predator to control aphids. The first use of a ladybird for biological control was the introduction of the 11-spot, *Coccinella undecimpunctata*, from Britain to New Zealand in 1874 to control aphid pests. The desirability of increasing the numbers of ladybirds was pointed out by Kirby & Spence as early as 1815. They suggested that if it were possible to induce the common English ladybird to proliferate, ornamental hot houses could be cleared of aphids. However, this suggestion has only recently been acted upon, mainly as a consequence of the need for more environmentally friendly ways of controlling insect pests in glass houses, orchards and gardens.

Until the third quarter of the 19th century none of the intercontinental transfers of predators or parasites was successful. The first great success was obtained in 1889 with the introduction of an Australian ladybird (*Rodolia cardinalis*) into California to control a scale insect (*Icerya purchasi*), which was devastating the Californian citrus industry. The head of the Entomological Bureau of the United States, C. V. Riley, instigated the introduction, but it was Albert Koebele, his assistant, who did the travelling and collecting. The success of this introduction was so immediate and complete that it aroused great enthusiasm and focused attention upon the possibilities of biological control of insect pests and became the classic case of biological control. *Rodolia* is still being used successfully against mealybugs. For example, *R. limbata* has recently been very effective in reducing the abundance of *I. aegyptiaca* destroying breadfruit trees on the coral atolls of Kiribati, Palau, the Marshall Islands and the Federated States of Micronesia in the Pacific. Within a short time of its release on an island the numbers of the ladybird dramatically increase and those of the mealybug decline, after which very low numbers of both remain on the trees in equilibrium (Fig.9.1; Veronica Brancatini & Don Sands, personal communication).

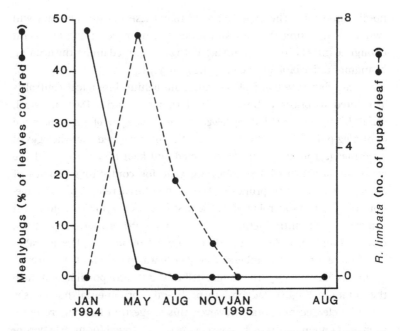

Fig. 9.1. The decline in the abundance of the mealybug *Icerya purchasi* (●——●) following the release of *Rodolia limbata* (● --- ●) on an atoll in the Federated States of Micronesia. (After Brancatini & Sands, personal communication.)

CONFLICT OF INTEREST

Specificity is seen as an important feature of a biological control agent and candidate species are intensively screened and the risk of their attacking non-target species assessed before their use in biological control programmes. However, questions about the safety of classical bio-logical control have been raised, particularly by conservationists con-cerned about the preservation of native flora and fauna, especially in New Zealand, Australia and Hawaii (Pimentel *et al.*, 1984; Longworth, 1987; Howarth, 1991). In spite of several claims that the introduction of exotic insect parasitoids and predators has caused the extinction or had a serious adverse affect on the abundance of an endemic species (e.g. Howarth, 1983, 1991) there is little empirical support for this suggestion (Hopper, 1998; Messing & Duan, 1998). This issue has initiated an ongoing debate between conservationists and biological control practitioners (Simberloff & Stiling, 1996, 1998; Frank, 1998), which because of the strongly held views is likely to continue for some time. In favour of biolog-ical control are the many successes, and the few outstanding cases of highly successful control like that of the use of *Rodolia* against *Icerya*. In addition, the most likely alternative, chemical control, is undoubtedly considerably more damaging to the environment.

Theory

From a theoretical point of view the successful introduction of a parasitoid or predator could have an adverse affect on the abundance of non-target insects, especially if the natural enemies are not specific to the target insect. As a pest insect by definition is likely to be abundant the large numbers of natural enemies developing on the pest could have a marked adverse affect on the abundance of rarer non-target endemic insects. This phenomenon is referred to as apparent competition (Holt, 1977). It is also possible that the introduced biological control agent might compete with endemic natural enemies. That is, theory predicts that the biological control of insects could have an adverse effect on the abundance of endemic insects by direct and apparent competition. What is difficult to predict is the order of the effect and whether it will be worse than the consequences of the side-effects of chemical control.

Apparent competition

The potential importance of apparent competition in community ecology has been discussed by Holt & Lawton (1994) and it is thought that communities of herbivorous arthropods might be structured by apparent competition mediated by shared predators and parasitoids (Jeffries & Lawton, 1984; Lawton, 1986; Godfray, 1994). Although the evidence that herbivorous arthropods interact through apparent competition is sparse, there is one field test of this idea involving a ladybird.

Müller & Godfray (1997) monitored the trends in the numbers of aphids (*Microlophium carnosum*) on nettle (*Urtica dioica*) placed around the perimeters of field plots of grass lightly or heavily infested with a grass aphid (*Rhopalosiphum padi*). The grass plots heavily infested with aphids attracted more adult seven-spot (*Coccinella septempunctata*) ladybirds than the lightly infested grass plots. The nettle aphid colonies (*Microlophium evansi*) adjacent to the heavily infested grass plots (experimental) suffered an earlier decline and produced fewer alate dispersers than the colonies adjacent to the lightly infested grass plots (control). The poorer performance of the nettle aphid colonies adjacent to the heavily infested grass plots was attributed to predation by the high numbers of ladybirds attracted to the area by the abundance of grass aphids. The numbers of ladybird larvae indicate that the ladybirds laid more eggs in the nettle aphid colonies adjacent to the heavily infested than the lightly infested plots, which gave rise to average peaks of 22 and 16 larvae, respectively. However, in spite of the higher abundance of nettle aphids adjacent to the lightly infested plots, especially towards the end of the experiment, only one ladybird pupa was recorded in these colonies, whereas an average of eight pupae were recorded from the lower-density nettle aphid colonies adjacent to the heavily infested plots (Fig. 9.2). This is the reverse of what one would expect. That is, although the results clearly indicate that lady-birds attracted to high concentrations of an aphid reduce the abundance of another rarer species of aphid living in the same locality, the virtual lack of ladybird pupae in the colonies of nettle aphid adjacent to the lightly infested plots raises questions about the mechanisms by which this was brought about.

Competition

Several species of ladybirds have been introduced into North America for biological control purposes and have become widely established and are extending their ranges. Assessing their effect on the abundance of indigenous species of ladybird is difficult because little if anything is known about their long-term fluctuations in abundance and because of the short time-scale over which such studies are usually done. In the extreme case it is claimed that species like *C. septempunctata* has already displaced the native *C. novemnotata* from Maryland nurseries (Staines *et al.*, 1990) and in parts of southern Africa the highly invasive *Chilocorus nigritus* is thought to have displaced the native *C. wahlbergi* (Samways, 1994). The introduction of *Rodolia cardinalis* into India is thought to have resulted in the displacement of the indigenous *R. amabilis* from colonies of cottony-cushion scale, *Icerya purchasi* (Subramanian, 1953). In New Zealand the native orange-spotted ladybird, *Coccinella leonina*, is similar in size and ecology to the introduced 11-spotted lady-bird, *C. undecimpunctata*. They both occur in the South Island and in the south of the North Island, but only the 11-spotted ladybird is present in the north of the North Island. Although absent from the mainland the orange-spotted ladybird is present on some of the northern offshore islands where the 11-spotted ladybird is absent. It is thought that the 11-spotted ladybird has a competitive edge over and displaces the orange-spotted ladybird in the warmer parts of the North Island, whereas they coexist further south where it is cooler (Watt, 1986). Another study of ladybirds, in the Auckland area of North Island, New Zealand, has

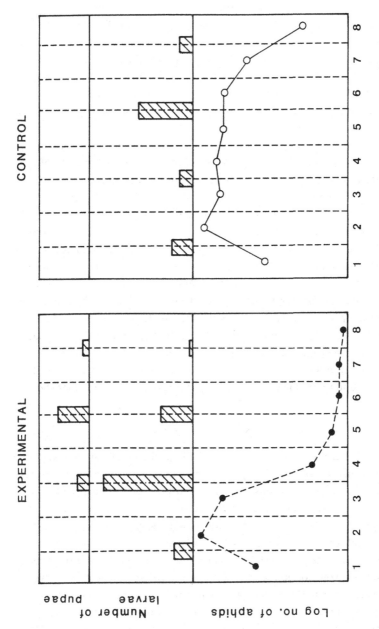

Fig. 9.2. Trends in the numbers of aphids, and the numbers of larvae and pupae of *Coccinella septempunctata* on nettles adjacent to heavily (experimental) and lightly (control) aphid-infested plots of grass. (After Müller & Godfray, 1997.)

recorded the distribution of species in those genera that have native as well as introduced species. The native species are only found in areas where the native vegetation still thrives, whereas the introduced species are mainly found in cultivated areas (Kuschel, 1990). Thus the disappearance of native species may be more a consequence of the destruction of their habitats than displacement by a superior competitor.

The more detailed study of Elliot *et al.* (1996) indicates that of the seven species of ladybirds that inhabited agricultural landscapes in eastern South Dakota prior to the invasion of the region by the introduced *C. septempunctata*, two species, *A. bipunctata* and *C. transversoguttata richardsoni*, are both some 20 to 30 times less abundant in certain crops than they were prior to the invasion. Another long-term study details the effect on 11 species of indigenous ladybirds of the arrival and establishment of *Harmonia axyridis* in Michigan (Calunga-Garcia & Gage, 1999). The effect was assessed by comparing the trends in abundance of the indigenous ladybirds for 5 years before and after the arrival of *H. axyridis*. Although *H. axyridis* became a dominant species, in terms of abundance, in a wide range of habitats, there is no evidence that it has had a marked negative affect on the abundance of any of the native ladybirds. The widespread concern about the adverse effects of such introductions on the abundance of native ladybirds justifies further research. However, it should be long-term and similar in design to that done by Calunga-Garcia & Gage (1999). If such introductions are shown to have an adverse effect then it would be interesting to know the processes by which the displacement and reduction in abundance of the native ladybirds comes about.

BIOLOGICAL CONTROL AND CONSERVATION

There is some evidence that biological control can be favourable for conservation where a pest is causing damage in a nature reserve (Samways, 1994). It has also been used with the express purpose of conservation. In the early 1940s two scale insects, *Carulaspis minima* and *Lepidosaphes newsteadi*, invaded the island of Bermuda and devastated the natural forests of Bermuda cedar (*Juniperus bermudiana*). Several natural enemies, including the ladybird *Rhizobius lophanthae*, were introduced in an endeavour to reduce the abundance of the scale insects and save the forests. In 1951 the project was abandoned, when most of the trees were dead or dying. Although the Bermuda cedar did not become extinct the role of the natural enemies in this is unknown (Bedford, 1949). More recently the scale insect *Orthezia insignis* invaded St Helena and nearly completely killed off the stands of the national tree, an endemic

Fig. 9.3. *Hyperaspis pantherina*.

gumwood belonging to the Compositae, *Commidendrum robustum*. The scale was successfully controlled and the gumwood saved by the introduction of the ladybird *Hyperaspis pantherina* (Fig. 9.3) by CABI Bioscience in 1993 (I am indebted to Marc Kenis for bringing this example to my attention and to Garry Hill for the details).

BIOLOGICAL CONTROL AGENTS

Although one of the aims of the study of population dynamics is to convert the art of biological control into a science it has so far contributed little if anything to the practice of biological control. This is particularly surprising bearing in mind the enormous effort that has been expended in the theoretical study of population dynamics. However, until the theoretical foundations of biological control are secure applied biologists can hardly be blamed for not using ecological principles and population theory when planning control measures against a pest.

Population theorists usually assume that herbivore abundance is regulated by the activity of natural enemies, which are specialists and have generation times similar to those of their prey. In fact the theory has been developed specifically for the interaction between parasitoids and their hosts. However, the first and great biological control success involved the use of a predator – *Rodolia*. This precedent resulted in the ladybird fantasy, and the world-wide and mainly ineffectual use of ladybirds as biological control agents. The success of *Rodolia* has undoubtedly

Table 9.1. *Attributes of biocontrol agents indicated by empirical and theoretical studies*

Empirical	Theoretical
1 Ecological capability	1 Synchrony or slight asynchrony
2 Temporal synchronization	2 High relative rate of increase
3 Density responsiveness	3 High searching efficiency
4 Reproductive potential	4 Interference amongst the natural enemies
5 Searching capacity	5 Aggregation on host patches
6 Dispersal capacity	6 Dispersal ability
7 Host/prey specificity and compatibility	
8 Food requirements	
9 Habitat requirements	
10 Natural enemies	

made it difficult for theorists to ignore predators completely. This is possibly reflected in the attempt to make the theory more general and all embracing by referring to it as the theory of predator–prey interactions. The justification for this appears to be that parasitoids and predators are essentially the same, and that as the parasitoids have simpler life histories it is easier to model their dynamics than the interaction between predators and their prey. However, the notion that the theory is relevant to predator–prey interactions is sustained by frequent reference to the similarities in the foraging behaviour of predators and parasitoids.

This led to the development of analytical models (p. 155) the parameters of which should reveal the attributes of effective biological control agents. It is not unusual for books on biological control to indicate either implicitly, or more rarely explicitly, the desirable attributes of biological control agents, usually with the claim that the selection has in some way been based on ecological principles. In a recent treatise on the subject (Van Driesche & Bellows, 1996) the attributes cited are those the authors regarded notable in species *used* in successful biological control programmes.

The analytical models similarly indicate a list of attributes (Table 9.1; Luck, 1990). That is, both approaches result in lists, which because of their generality provide little guidance to the practitioners.

The theoretical studies have mainly contributed to our understanding of the factors that stabilize an otherwise unstable interaction. For the practice of biological control the theory has only shown that it is possibly misguided to oppose the introduction of additional parasitoids due to concerns about potential competition between parasitoids adversely affecting the existing level of biological control (Hassell & Varley, 1969; May & Hassell, 1981). However, even on this point Murdoch *et al.* (1998) suggest we ask 'What kind(s) of species, and how many, should we release given the modes of competition and coexistence among them?' rather than 'How many species should we release?' That is, there is still no clear directive from the theoreticians.

The practitioners have found it difficult to define the exact nature of the attributes listed above and to measure them in the field, particularly if different species are to be compared (Waage & Mills, 1992). This has led to the proposal that the important, above all easily measurable, life history parameters, which can be collected in a few months of field or laboratory study, should be incorporated into a prospective model. The parameters included in one such model were: stage of host attacked by different parasitoids, age-specific development rates for host and parasitoids, age-specific survivorship of hosts in the field, and adult longevity and daily oviposition rate. The difficult-to-measure parameters such as searching efficiency are incorporated into the model by treating them as variables. This has the virtue of being directed at a specific pest problem and is a more practical approach when the model must be built, and used quickly and cheaply (Waage & Mills, 1992) Although the predictions of such models are supported by field observations, there are discrepancies, which caution against unconsidered extrapolation of theoretical predictions to specific situations (Gutierrez *et al.*, 1994). Thus, although these models appear to reveal that there is a good understanding of the nature of all the important interactions, they do not reveal what it is about the natural enemies that make them successful in reducing pest abundance. The challenge is to use these models (elegant specific descriptions) to forward the practice of biological control.

LADYBIRDS AND BIOLOGICAL CONTROL

Although biological control has contributed greatly to our understanding of coevolution (myxomatosis), plant distribution (St John's wort) and competition (*Aphytis*) it has as indicated above not contributed much to the practice of biological control. This has led several authors to recommend that more attention be given to studying well-documented cases of successful and unsuccessful attempts at classical biological control (Luck, 1990; Kareiva, 1996). That is, there is a need to diagnose the sources of failure and success. A serious limitation to this approach is that there has only been one rigorous field study of the interaction between introduced natural enemies and pest in a successful (Roland, 1998) and no field study of an unsuccessful biological control programme. However, the idea has considerable merit. Another possible approach is to study a group, like ladybirds, whose use in biological control has resulted in the classical case and many failures. Selecting ladybirds has the additional merit of reducing the phylogenetic problem in that they are all members of the same family of beetles, the Coccinellidae, and their general biology and that of their prey has been very well studied. This is done here by examining the data in the literature for evidence of patterns, and then the possible processes shaping these patterns are analysed with the objective of identifying specific attributes for success.

World-wide there have been 155 attempts to control aphids and 613 to control coccids by introducing ladybirds. The outcome of each introduction is classified as either complete, substantial, partial or no control. Of the attempts to control aphids only one was substantially successful, which is significantly less than the 53 complete and substantial successes achieved against coccids. In terms of complete control there have been no successes against aphids whereas 23 of the attempts to control coccids were completely successful. That is, the outcomes of the many attempts to use ladybirds to reduce the abundance of pest aphids and coccids in the field clearly indicate that coccidophagous ladybirds are more effective than aphidophagous ladybirds (Dixon & Kindlmann, 1998).

Attributes of successful biological control agents

The attributes of good biological control ladybirds and the reasons for the failures and successes of ladybirds to reduce the abundance of a pest cited in the literature are given in Table 9.2. Some of the attributes and reasons, such as 'abundance of the predator' and 'ephemeral nature of aphid populations' are a consequence of the operation of other processes. Problems created by a mismatch in the temperature thresholds of predator and prey (cf. p. 3) are now usually overcome by climatic mapping (Samways, 1989). Others like 'attacks all stages of prey', which has also been highlighted as important by theoreticians (Murdoch, 1990; Lane *et al.*, 1999), apply to both aphidophagous and coccidophagous ladybirds. However, the fact that several of the very successful coccidophagous ladybirds appear to have a greater preference for particular developmental stages of their prey than aphidophagous species tends to cast doubt on the importance of this attribute. Most biological control practitioners consider searching ability/capacity to be one of the most important attributes of an effective natural enemy (Nechols & Obrycki, 1989).

Table 9.2. *Attributes of ladybirds that are thought to make them good biological control agents, and the reasons given for the failure of aphidophagous ladybirds and the success of some coccidophagous ladybirds in biological control programmes*

Attributes of biological control agents	Reasons for:	
	Failure of aphidophagous ladybirds	Success of coccidophagous ladybirds
Abundance of predator (Frazer, 1988; Hagen, 1974; Hodek, 1967)	Hagen (1974) Ephemeral nature of aphid populations	Thorpe (1930) Largely independent of climatic conditions
Voracity (Mills, 1982a; Taylor, 1935; Thorpe, 1930)	Competition with endemic natural enemies	Specific and voracious
Searching ability	Lack of synchrony	Attack all stages of prey
Temperature thresholds	Absence of suitable shelter sites	High searching efficiency
Relative rate of increase	Mills (1982) Satiated at moderate to high aphid densities	High relative rate of increase
	Little prey specificity	Free of natural enemies
	Development slow relative to aphids	Smith (1939) Dispersion of host relative to the searching ability of predator

Although little work has been done on the searching efficiency of lady-birds, especially in the field, there is no indication that the two groups differ in this respect and from an evolutionary point of view it is difficult to see why they should.

On the question of searching ability it is interesting to note that Smith (1939) thought that the effectiveness of a predator is dependent on its power of discovery relative to the dispersion of the prey. He illustrated this by reference to two ladybirds that were introduced into California to control two species of scale insect. *Rodolia cardinalis* was, as indicated above, very successful in reducing the the abundance of cottony-cushion scale, but *Rhizobius ventralis* was not successful in controlling black scale. The adults of these two ladybirds appear to be very similar in their search-ing ability. In addition, the data in the more recent literature indicate that these two species are likely to have similar developmental times (Fig. 9.4). The success of *Rodolia cardinalis* Smith (1939) attributed to its larvae each completing their development on the contents of a large egg sac of a cottony-cushion scale, and the failure of *Rhizobius ventralis* to its larvae not being able to mature on one scale and having to search for other prey. Thus when the population of prey becomes scarce the larvae of *Rhizobius ventralis* are at a considerable disadvantage compared with those of *Rodolia cardinalis*. However, as the larvae of *Rhizobius ventralis* have rela-tively much longer legs than the larvae of *Rodolia cardinalis* (Fig. 9.4), they are likely to move faster, which might compensate for the difference in the dispersion of their prey. Interestingly, although Smith's study was published 60 years ago, it shows that a comparative analysis of successful and unsuccessful cases of biological control, as currently being advo-cated, is likely to highlight the attributes of potentially successful biolog-ical control agents. In addition, it indicates that it may not be the absolute value of an attribute of a predator that is important but its value relative to that of some feature of the prey.

Certain attributes are highlighted as important by several authors. These are: specificity, voracity and relative rate of increase. Below each of these are reviewed in detail.

Specificity

Although predators may be less specific in their food habits than parasitoids, the difference is not likely to be as great as we have been led to believe (Thompson, 1951). In fact, there are monophagous insect pred-ators, like the coccidophagous ladybird *Coelophora quadrivittata* that feeds only on *Coccus viridis* (Chazeau, 1981) and the aphidophagous chry-

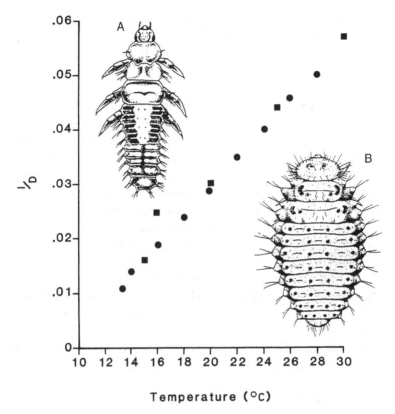

Fig. 9.4. The developmental rates of (A) *Rhizobius lophanthae* (■) and (B) *Rodolia cardinalis* (●) recorded at different temperatures. (After Bodenheimer, 1932; Cividanes & Gutierrez, 1996.)

sopid *Chrysopa slossonae* that feeds only on *Prociphilus tesselatus* (Bristow, 1988).

There is some indication, however, that specificity could be an imporatnt attribute of a biological control agent. The coccidophagous ladybirds that feed on the Margarodidae, that is the group of coccids that includes *Icerya*, are generally more specific than those that feed on other groups of coccids (Froggat, 1902 ; Kairo & Murphy, 1995). Of the 79 attempts to use ladybirds to control Margarodidae 15 have been completely successful, compared to 7 of the 534 attempts to use them to control other groups of coccids. That is, the proportion of complete successes is significantly greater for the more prey-specific coccidophagous ladybirds ($\chi^2 = 61.2$, $P < 0.001$). In addition, the polyphagous *Rhizobius* failed to control the black scale in California whereas the prey-specific *Rodolia* successfully controlled the cottony-cushion scale there, even though in other respects relative to their prey these two ladybirds appear to be very similar.

Voracity

Other things being equal predaceous insects would appear to have a greater power of destruction than parasitic insects and thus a greater potential importance in pest control (Thompson, 1951). Voracity is indeed considered an important feature of an effective biological control predator (Taylor, 1935; Gurney & Hussey, 1970; Samways & Wilson, 1988). However, the potential voracity of a population of predators is limited by their long generation time relative to that of their prey and by the fact that increasing the number of predators produces an additive increase in potential voracity, whereas the numbers of the prey increase geometrically. Bombosch (1963) developed a simple model in which predator voracity is balanced against the geometric increase of the prey. This and a subsequent study by van Emden (1966) using the same model indicated that the main factors determining the effectiveness of a predator are voracity (a function of appetite, activity and abundance), synchronization with the prey and the multiplication rate of the prey. That is, the faster a prey reproduces, the more voracious must a predator be.

Attack rate

Others have favoured the use of attack rate (p. 82) rather than consumption rate because prey may die after a predator attack without being consumed (Hodek *et al.*, 1984a; Stäubli Dreyer *et al.*, 1997a). This can be obtained from the functional response equation :

$$Na = a N / (1 + a\, Th\, N) \tag{9.1}$$

where Na is the number of prey attacked per predator per day, Th the handling time associated with each prey attacked, N the density of prey and a the instantaneous attack rate (Holling, 1959). A modified version of this approach has been used by Baumgaertner *et al.* (1987), Gutierrez *et al.* (1981) and Cividanes & Gutierrez (1996). In this

$$Na = W(a)\, g(.) = W(a)\, D\, [\, 1 - \exp(\alpha\, mS / D\, W(a))] \tag{9.2}$$

where W is dry weight of a predator of age a, $g(.)$ the per unit mass functional response, D the maximum demand rate for prey, α the proportion of prey that may be attacked and S the number of prey of average dry mass, m. That is, the instantaneous attack rate has been replaced by demand rate and expressed in terms of consumption (mg/mg/day) rather than numbers. There are no studies in which the demand rates of coccidophagous and aphidophagous ladybirds have been compared. However, even if one accepts that functional responses are meaningful in this context (cf. p. 166), then assuming that the proportion of prey consumed is independent of prey type, the attack/demand rates are likely to indicate that aphidophagous ladybirds are more voracious than coccidophagous species.

Conversion efficiency

Coccidophagous species of ladybirds are reported to feed almost continuously, whereas aphidophagous species are characterized by long periods of inactivity due to satiation (Taylor, 1935; Mills, 1982b). In feeding on immobile prey, coccidophagous ladybirds are thought to be more selective, concentrating on that part of their prey that is readily extracted, easily assimilated and rapidly digested. A reduction in the digestion time would allow more continuous feeding, while a reduction in the proportion of each prey item eaten would increase the potential impact on the prey population. In contrast aphidophagous species tend to eat all of each prey item, and this non-selective feeding results in frequent satiation. This led Mills (1982b) to suggest that the great success of coccidophagous species in biological control is attributable to their more optimal use of prey. The facts that the rate of development of coccidophagous species is slower than that of aphidophagous species (Fig. 4.2; p. 70) and the efficiencies of converting prey biomass into ladybird biomass are very similar in these two groups of ladybirds (Table 9.3) do not support the above contention. However, Mills's optimum food utilization/satiation hypothesis has not been specifically tested. Current studies, however, indicate little if any difference in the feeding behaviour or the proportion of time spent inactive in these two groups of ladybirds (Magro, A., personal communication).

If voracity is all-important in determining whether a ladybird is likely to be an effective biological control agent then coccidophagous species should be more voracious than aphidophagous species. Most measures of voracity in the literature consist of the number of prey eaten per unit time or per developmental stage. However, predators vary in size and prey vary greatly in size both between species and between developmental stages within a species. Therefore, such information is of very little value for comparing the voracity of different species of predator. A way of resolving this problem is first to determine the conversion efficiency, i.e. express the adult weight of a predator as a proportion of the total weight of prey it needs to complete its development. This overcomes

Table 9.3. *The efficiency with which six aphidophagous and four coccidophagous species of ladybird convert prey into ladybird biomass*

Species	Conversion efficiency = adult weight (μg)/ larval consumption (μg)	Source
Aphidophagous		
Adalia bipunctata	0.15	Blackman (1967)
	0.30	Ferran *et al.* (1984)
	0.1–0.17	Mills (1979)
Coccinella septempunctata	0.13	Formusoh & Wilde (1993); Wetzel *et al.* (1982)
Harmonia axyridis	0.29	Schanderl *et al.* (1985)
Olla v-nigrum	0.14	Kreiter & Iperti (1984); J.-L. Hemptinne, pers. comm.
Propylea quatuordecimpunctata	0.30	Quilici (1981)
Semiadalia undecimnotata	0.24	Ferran & Larroque (1979)
	0.28	Ferran & Larroque (1979)
Coccidophagous		
Diomus hennesseyi	0.24	Kanika-Kiamfu *et al.* (1992)
Exochomus flaviventris	0.12	Kanika-Kiamfu *et al.* (1992)
Hyperaspis raynevali	0.29	Kanika-Kiamfu *et al.* (1992)
Rhizobius lophanthae	0.14	Cividanes & Gutierrez (1996)

the confounding effects of the relative sizes of the predator and prey. There are very few data sets in the literature that can be used for making this calculation. However, the range of values for six species of aphidophagous ladybirds is similar to that for four species of coccidophagous ladybirds (Table 9.3) On the basis of this very small sample it is unlikely the conversion efficiencies of these two groups of ladybirds differ. The above values can be converted to rates by dividing by the time it takes each species to complete its development. As the immature stages of coccidophagous species take considerably longer to develop to maturity than aphidophagous species the larval consumption of the latter expressed as weight of prey per unit weight of adult per unit time will clearly be much larger than that for coccidophagous species.

Relative growth rate

More simply the information used to obtain the above values can also be used to calculate the relative growth rates of the ladybirds. Although there are very few data sets that are detailed enough, nevertheless the relative growth rates for the two groups of ladybirds clearly support the above conclusions (Table 4.1). That is, aphidophagous ladybirds grow faster and therefore not surprisingly consume more prey per unit time per unit weight than coccidophagous species. This is the reverse of what one would expect if voracity were an important attribute of a biological control agent. In addition varying voracity in the model developed specifically for the ladybird–prey interaction (p. 160; Kindlmann & Dixon, 1993) revealed that voracity had very little effect on the overall abundance of the prey compared to varying the relative generation times of the predator and prey.

Fecundity

Predators generally kill more than one prey individual during the course of their development, whereas parasitoids only kill one host. The greater killing power of predators has been seen as a positive attribute in biological control agents (Thompson, 1951). In the case of parasitoids this line of thinking is reflected in the belief that all other things being equal species with a higher fecundity are likely to be better potential biological control agents because of their ability to kill a greater number of hosts (Lane et al., 1999). That is, they are likely to be able to reproduce more rapidly than the pest and respond to changes in its abundance. Similarly, the potential killing power of predators might also be seen to be dependent on both their voracity and lifetime fecundity. However, the coccidophagous ladybirds, which have been the more successful biological control agents, have a lower lifetime fecundity than the less successful aphidophagous ladybirds (p. 74). The poor performance as biological control agents of some parasitoids with high fecundities has been attributed to trade-offs between searching efficiency or longevity against fecundity (Lane et al., 1999). In ladybirds there also seems to be a trade-off between longevity and fecundity (p. 79). The implication is that the greater success of coccidophagous ladybirds as biological control agents is attributable to their greater adult longevity. However, the longer adult life and lower fecundity of coccidophagous ladybirds may not constitute a trade-off but rather a further expression of their pace of life relative to that of aphidophagous species. Coccidophagous species generally do everything more slowly than aphidophagous species. That is, the parameters that make up the suites of life history parameters characteristic of aphidophagous and coccidophagous ladybirds are in each case very closely linked with little if any trade-off between them (p. 80).

In summary, however one measures voracity it does not appear to be the attribute that accounts for the greater success of coccidophagous ladybirds in biological control. In absolute terms coccidophagous ladybirds are less voracious than aphidophagous species.

Generation time ratio

Several authors have noted that the rates of development of predators and prey can differ, and that when a predator develops slower than its prey it is an ineffective biological control agent (Bombosch, 1963; Hagen, 1974; Mills, 1982b) and when faster or at the same speed an effective biological control agent (Thorpe, 1930; Taylor, 1935). However, relative developmental rate is presented along with a list of other attributes with no indication of its relative importance. As under optimum conditions there is a very close association between developmental rate $(1/D)$ and intrinsic rate of population increase (r_m) (Fig. 7.10) the ratio of the developmental times of a predator and its prey could be the most important single factor determining the abundance of the prey (cf. Fig. 7.11).

The developmental time of aphidophagous ladybirds often spans several aphid generations (Figs. 4.2, 4.3), so the ratio of generation time of these predators to that of their prey is large and a ladybird's rate of increase depends not only on the present state of a patch of prey, but also on the quality of the patch in the future. Its rate of increase is therefore mainly determined by expectations of future bottlenecks in prey abundance. In the case of the coccid ladybird interaction the generation time ratio is 1.0 or less. That is, bottlenecks in prey abundance are not a problem for coccidophagous ladybirds.

The average reduction in prey density below the predator-free value (q) (Beddington et al., 1978) calculated using a model incorporating the above concepts (p. 168) and different generation time ratios are given in Fig. 9.5. This indicates that an effective biological control agent must develop faster or at a rate similar to that of its prey/host. The empirical data for aphidophagous and coccidophagous ladybirds supports the prediction.

Ladybirds are not only predators of aphids and coccids. Some prey exclusively on other groups of arthropods. Although these ladybirds have not been as well studied as the aphidophagous and coccidophagous species, there are sufficient data in the literature to indicate their rates of development and that of their prey. The relationship between the devel-

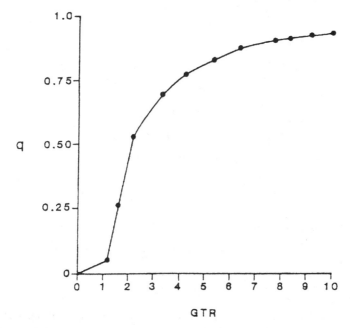

Fig. 9.5. The relationship between the average reduction in prey density below the predator-free value (*q*) for different generation time ratios of predator and prey (GTR) predicted by the Kindlmann & Dixon (1999) model.

opmental rate of each group of ladybirds and that of their prey (Fig.9.6) is curvilinear, which indicates that there is an upper bound to the rate of development in ladybirds and in most cases ladybirds develop more slowly than their prey. The marked exceptions are the coccidophagous species many of which are known to be very effective biological control agents. Interestingly, the ladybird predators of whitefly (aleyrodids) and psyllids have developmental times similar to those of their prey. Although these particular ladybirds have been far less frequently used in biological control programmes they have been claimed to be effective biological control agents (Smith & Maltby, 1964; Leeper & Beardsley, 1976; Kumashiro *et al.*, 1983; Hoelmer *et al.*, 1993; Heinz *et al.*, 1994). Similarly the ladybirds that feed on adelgids or mites, which have generation time ratios that are only slightly bigger than 1.0, are also claimed to be effective at reducing the abundance of their prey in the field (Sasaji & McClure, 1997). The aphidophagous ladybirds, relative to their prey, have the largest generation time ratio and the ineffectiveness of these ladybirds is well documented.

Although in general terms the most successful biological control ladybirds are likely to have similar or shorter developmental times than

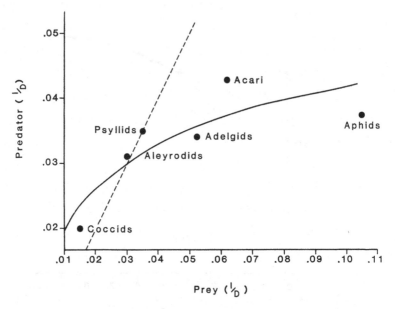

Fig. 9.6. The relationship between the average developmental rates (1/D) of acari, adelgids, aleyrodids, aphids, coccids and psyllids, and that of their specific ladybird predators at 20 °C. The dashed line indicates a developmental rate of the predator divided by that of its prey (GTR) equal to 1.0.

their prey not all attempts to control coccids using ladybirds have been successful (p. 200). This indicates that success is determined by more than just the generation time ratio. After comparing the success of *Rodolia* in controlling the cottony-cushion scale with the failure of *Rhizobius* to control the black scale Smith (1939) attributed the success of *Rodolia* to its greater powers of discovery relative to the dispersion of its prey (p. 202). However, there are grounds for thinking that *Rhizobius* may be as well adapted to exploiting its prey as *Rodolia*. At 25 °C the black scale's rate of development is 1.5 times faster than that of the cottony-cushion scale, but nevertheless, both predators develop faster than their prey, the more so in the case of *Rodolia*. The literature indicates that these two predators differ in another respect: *Rhizobius* is polyphagous (Cochereau, 1969), and *Rodolia* is highly prey-specific (Froggat, 1902; Kairo & Murphy, 1995). Thus prey specificity, as indicated above, could also be an important attribute of a successful biological control agent. A more detailed analysis of the successes and failures of ladybirds to control coccids is needed to specifically test this suggestion.

In summary, of the many attributes of predators that are claimed to be important in determining their effectiveness as biocontrol agents only two clearly differ in aphidophagous and coccidophagous ladybirds; prey specificity and generation time ratio. A simple model has shown that prey abundance is very sensitive to generation time ratio. The case for prey specificity being important needs to be more clearly established and its precise effect on prey abundance remains to be shown.

AUGMENTATIVE BIOLOGICAL CONTROL

Ladybirds have been widely used to control aphids and coccids through augmentation by translocation or mass rearing and release. The translocation of *Coccinella septempunctata* from wheat to cotton is a strategy widely used to control the cotton aphid, *Aphis gossypii*, in central and southern China. In May, when *C. septempunctata* is abundant ($9-12/m^2$) in wheat fields in central China, large numbers are collected and released in cotton fields. A release rate of $1.5-3/m^2$ reduces aphid abundance by 98% in 2 days (Pu, 1976). Mass rearing and release of *Cryptolaemus montrouzieri* has similarly proved an effective means of controlling outbreaks of the mealybug *Pseudococcus citri* on citrus in California (Smith & Armitage, 1920).

The use of ladybirds in augmentative biological control has been hampered by the tendency of the adults to fly away. For example, between 1908 and 1914 huge numbers of hibernating adults of *Hippodamia convergens* were transferred from overwintering sites in the Sierra Nevada mountains of California to cantaloup growers in the Imperial Valley in an attempt to control melon aphid, *Aphis gossypii* (Carness, 1912*a*, *b*). More recently ladybirds have been used to control *A. gossypii* on chrysanthemums and *Macrosiphum rosae* on rose bushes (Flint *et al.*, 1995; Dreistadt & Flint, 1996). However, the beetles rapidly disperse and scarcely any remain in the release areas after 3 days (Hagen, 1962; Flint *et al.*, 1995; Dreistadt & Flint, 1996). To overcome this larvae and flightless adults are now used.

The mass production of ladybirds is dependent on the availability of large quantities of prey, which can be difficult to maintain. In an endeavour to overcome this there have been many attempts to rear ladybirds on artificial diets (Smith, 1966*b*; Matsuka *et al.*, 1982; Hattingh & Samways, 1993). *Harmonia axyridis* was introduced into France in 1982 and since then has been reared continuously for over 100 generations on industrially produced eggs of the moth *Ephestia kuehniella* (Schanderl *et al.*, 1985, 1988). The larvae now only show a weak response to the presence of aphids and their capture efficiency is reduced compared to larvae of beetles that have been reared continuously on a diet of aphids. This reduced response to aphids could adversely affect their effectiveness as biological control

agents. Although the effect of continuous rearing on moth eggs is temporary it may account for why a large proportion of *H. axyridis* larvae
released on rose bushes infested with the rose aphid, *M. rosae,* disappeared
within a few hours of their release (Ferran *et al.,* 1997b). However,
advanced larvae (third and fourth instars) of *H. axyridis* have been used
effectively against the rose aphid (*M. rosae*) and damson-hop aphid
(*Phorodon humuli*) (Trouve *et al.,* 1997; Ferran *et al.,* 1998) and of *Adalia
bipunctata* against the rosy apple aphid (*Dysaphis plantaginea*) (Wyss *et al.,*
1999). In addition augmentative releases of adults of *Hippodamia convergens* have been used to reduce the abundance of aphids on ornamental
shrubs (Raupp *et al.,* 1994).

Forty years ago it was proposed that the effectiveness of biological
control agents might be improved by selection (DeBach, 1958; Sailer,
1961). As adult ladybirds will leave crops heavily infested with pests
Ferran and his colleagues at Antibes, France, have been attempting to
produce flightless adults, which by remaining longer on a crop would
provide the potential for more long-term control (Marples *et al.,* 1993).
Initially young males of *Harmonia axyridis,* treated with a mutagen, were
mated with virgin females and their offspring, which as adults had wing
malformations, were selected for breeding. The frequency of abnormal
adults (open elytra and extended wings) increased up to the eighth generation and then varied between 60 and 90%. Although their poor survival
and low fecundity made them poor candidates for mass rearing for biological control, nevertheless, when released in greenhouses containing
cucumber infested with *Aphis gossypii* they remained on the plants in
higher numbers and laid eggs over a longer period than the control
adults, but their larvae were less numerous (Ferran *et al.,* 1998). By selective breeding from a laboratory population of *H. axyridis* Tourniaire *et al.*
(2000b) have produced an homozygous flightless strain of this ladybird,
which has normal elytra and wings, and whose foraging behaviour and
fecundity is similar to that of normal beetles.

Cultural control

The rearing of large numbers of biological control agents for augmentative control programmes is expensive. Although increasing the
effectiveness of the biological control agents by selective breeding is
likely to make it more cost-effective it will continue to be costly and
subject to disruption by disease etc. To avoid this several researchers have
attempted to increase the abundance of native natural enemies in the
immediate vicinity of a crop. This approach has been championed by the

conservationists and caught the attention of the media. Hedgerows, unsprayed headlands, strips of flowers and beetle banks in promoting the abundance of natural enemies are seen by many as the most effective means of reducing the incidence of pests. Although it has attracted a lot of attention recently this methodology is not new as it was being advocated and tested 70–100 years ago (Sanborn, 1906; Marcovitch, 1935). However, the effectiveness of very few of these manipulations has been rigorously assessed.

The productivity of pecan orchards in the U.S.A. is seriously threatened by several species of leaf-feeding aphids. Bugg *et al.* (1991) and Bugg & Dutcher (1993) planted cover crops under pecan trees with the intention of attracting in natural enemies and testing the idea that this would give improved biological control of the pecan aphids. Manipulating the cool season cover crop gave a 6-fold, and the warm season cover crop a 48–125-fold increase in the abundance of ladybirds in the cover crop. However, this did not result in an increase in the abundance of the ladybirds or decrease in the abundance of the aphids on the pecan trees. Similarly, planting weed strips in apple orchards in Europe resulted in an increase in ladybirds etc. in the weed strips but not of ladybirds on the apple trees. In this case, however, the higher numbers of spiders in autumn on the trees underplanted with weeds reduced the abundance of the winged aphids returning to lay eggs and thus the number of aphids present the following spring (Wyss, 1995). The underplanting of maize with the weed *Acalypha ostryaefolia* provides an alternative oviposition site and refuge from cannibalism for *Coleomegilla maculata*, which results in an increase in the numbers of its larvae and of the mortality of the pest aphid on maize (Cottrell & Yeargan, 1999).

The many failures of cultural manipulations to increase the number of ladybirds on a particular crop is not surprising if one accepts that ladybirds and other biological control agents are only likely to stay, feed and lay eggs if it is advantageous in terms of their fitness (p. 102). One way round this fitness constraint is to encourage ladybirds etc. to breed on a catch crop and then when the larvae are in an advanced stage of development destroy the catch crop and so force the larvae to seek food on an adjacent cash crop. This has been very elegantly done by planting strips of faba bean (*Vicia faba* L.) between strips of lettuce. The faba beans often become heavily infested with black bean aphid, *Aphis fabae fabae*, and supply a breeding area for several ladybirds: *Adalia bipunctata*, *Coccinella septempunctata* and *Propylea quatuordecimpunctata*. Mowing the strips of faba beans forces the ladybird larvae to disperse and forage for food on adjacent lettuce plants. As a result the aphid abundance on lettuce between strips of faba beans was reduced by 22–83%. The extent of the reduction is dependent on the distance between strips. Release of marked larvae confirmed that they do disperse, and that this behaviour is well described by a diffusion model, which indicates it is only likely to be effective over a range of a few metres (Nunnenmacher, 1998).

In summary, because of their oviposition behaviour, increasing the numbers of adult ladybirds in a habitat will not necessarily result in a marked increase in the number of eggs laid in each patch of aphids. However, in habitats that unlike orchards are not vertically stratified large larvae may move between patches of aphids, especially if they are close by and differentially infested. That is, cultural manipulations are most likely to be successful in herbaceous crops, especially if the aphid infestations on weeds are declining when those on the crop are increasing.

INTEGRATED PEST MANAGEMENT

Biological control practitioners usually assume that natural enemies are capable of keeping pest populations below economic thresholds (Fig.9.7; Bonnemaison, 1966 ; Suter & Keller, 1990) but frequently fail to do so. This has stimulated a different approach to pest problems (Stern et al., 1959). 'Integrated control' or 'integrated pest management' utilizes all suitable techniques, including pesticides, to reduce the abundance of pests to below their economic thresholds. This is achieved by harmonizing several techniques, like host-plant resistance and cultivation, and the use of pesticides and natural enemies into a flexible system (Smith & Reynolds, 1972). However, should the pest become abundant then a pesticide is used. That is, providing the monitoring is effective there should be no further pest outbreaks.

As the system is programmed to deal with failures it is difficult to judge its success. One way would be by the quantity and/or toxicity of the pesticides used. As the abundance of the pest is continually being monitored any reduction in the application of pesticides could be more a consequence of eliminating prophylactic spraying than of having developed a system in which the biological components are more effective in reducing pest abundance. However, until the cost-effectiveness of such management practices has been rigorously assessed they are unlikely to attract universal acclaim.

Often an essential part of the management practice is the use of more pest-resistant varieties of crop. At first this appears be reasonable as it slows down the rate of development and rate of population increase of

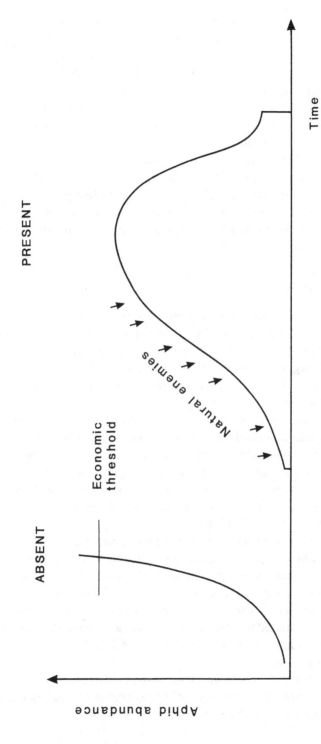

Fig. 9.7. The assumed trends in aphid abundance in the absence and presence of natural enemies. (After Suter & Keller, 1990.)

the pest. However, as many of the resistance mechanisms of plants involve secondary plant substances, and these are detoxified by the same enzymes that detoxify pesticides, this practice could be counterproductive as it results in the selection of pests that are also more likely to be resistant to pesticides.

In addition the enhancement of natural enemy activity is only likely to be effective if their combined effect on pest abundance is additive or synergistic (p. 175). The evidence from intraguild predation studies tends to indicate that the natural enemies are more likely to have a negative than a positive effect on one another. Also adult natural enemies are likely to avoid ovipositing in patches of a pest where predators or parasitoids are already present. That is, natural enemies are likely to respond to cues that optimize their fitness rather than indulge in behaviour that is optimal for pest control.

IS BIOLOGICAL CONTROL EVOLUTIONARILY STABLE?

In view of the widely reported evolution of resistance by insects and weeds to pesticides the virtual absence of reports of the evolution of resistance of pests to predators and parasitoids is surprising (Holt & Hochberg, 1997; Jervis, 1997). However, it presupposes that the evolutionary response of target pests to pesticidal and classical biological control involve similar constraints.

Pesticides in many respects are similar in their effect to secondary plant chemicals, and indeed some are just that. That is, they are part of a general defence and the ongoing coevolutionary arms race between herbivores and plants. Thus, it is not surprising that pests often develop resistance to these chemicals, using the same or similar detoxification systems as they use against secondary plant chemicals. That is, it involves the same basic mechanism.

In the case of classical biological control it is possible that the specialist natural enemies depend for their success on developing as fast as or faster than, the pest (Kindlmann & Dixon, 1998). It is likely that all pest organisms have been strongly selected to develop rapidly, i.e. there are no advantages for a pest in developing slowly and that development rate has been maximized by selection. If this is the case then the one way of escaping from natural enemy control – to develop faster – is not available or greatly limited.

In conclusion, ladybirds supplied the first case of classical biological control and continue to be successfully used against pests. However, this can result in a conflict of interest between conservationists and biological control practitioners. There are theoretical grounds and some empirical evidence to indicate that introduced species of ladybirds might adversely affect the abundance of rare species of prey and native species of ladybird. However, there is no good evidence of this having occurred and the most likely alternative to biological control, chemical control, is undoubtedly considerably more damaging to the environment. On the plus side ladybirds have also been used to control introduced pests ravaging endemic species of plant.

Coccidophagous ladybirds have been more successful biological control agents than aphidophagous species. This appears to be mainly a consequence of coccidophagous ladybirds developing faster than their prey and possibly also to their greater prey specificity. The effectiveness of augmentation is constrained by the cost of producing large numbers of ladybirds in the expectation of a pest outbreak and the effect of rearing ladybirds continuously on artificial food is likely to have on their foraging behaviour. Similarly the success of augmentation by cultural manipulations is also likely to be constrained by the behaviour of adults, which require very specific conditions for oviposition. The supposed effectiveness of integrated pest management is more likely a consequence of a closer monitoring of pest populations than manipulation of natural enemies. The long-term stability of biological control is possibly due to the fact that it is based on relative developmental rates, which have been maximized by selection.

10
Epilogue

Studies of insect predator–prey dynamics have been neglected in favour of those on parasitoid–host dynamics. This can be justified as parasitoids are more effective biological control agents than predators. The study of parasitoid–host dynamics, therefore, is seen as the most likely to yield a general theory for biological control practices. However, this approach has not resulted in useful recipes for manipulating pest populations in the field.

Although aphidophagous ladybirds have a poor sensory capability they appear to forage in a way that is similar to that predicted by optimal foraging theory. Adults by responding to cues that indicate the presence of aphids and conspecifics can select high-quality patches of prey for their offspring and so maximize their fitness. Similarly, hunger-induced changes in the searching behaviour of larvae serve to keep them in a patch close to clumps of prey, which enables them to maximize their rate of energy intake. That adults lay more than the optimum number of eggs per patch, however, is possibly the best strategy as the future quality of a patch is very uncertain. Future studies need to test these ideas and determine whether they apply to other ladybirds and insect predators.

Predaceous ladybirds differ in their speed of life, which is positively associated with that of their prey. Coccidophagous ladybirds all tend to develop, reproduce and age more slowly than aphidophagous ladybirds. That is, one easily measured life history parameter can be used as an index of all the other life history parameters, as there appears to be little or no trade-off between them. This interdependence of life history parameters offers a way round the daunting complexity that has resulted from the traditional approach to predator–prey dynamics. The resulting simplification of the interactions reveals that the rate of development of a predator relative to that of its prey could be important in determining whether a predator can reduce the abundance of a pest to non-damaging levels of abundance. An analysis of the important attributes of successful biological control agents cited in the literature also reveals the interdependence of life history parameters and adds further support for the notion that the most important attribute is the relative rate of development of predator and prey – the generation time ratio. The low prey specificity of some coccidophagous ladybirds possibly nullifies the effect of their low generation time ratio and accounts for their failure as biological control agents. That is, a low generation time ratio and high prey specificity appear to be the most important attributes of a good biological control predator. Both of these attributes have the added virtue of being easily measured.

There is increasing evidence that the relationships between ladybirds and their prey apply to other insect predators. If this is the case then it is likely that the patterns and processes revealed by this study of a well-researched group of insects are likely to apply generally to all insect predators. This book will have succeeded if it stimulates further research on ladybirds and other insect predators to test specifically the ideas presented here. In addition if the ideas prove to be general to both insect predators *and* parasitoids then in the future there will be less need to record the fall of every apple.

References

Agarwala, B. K. (1991) Why do ladybirds (Coleoptera, Coccinellidae) cannibalize? *Journal of Bioscience* **16**, 103–109.

Agarwala, B. K., Bhattacharya, S. & Bardhanroy, P. (1998) Who eats whose eggs? Intra- versus inter-specific interactions in starving ladybird beetles predaceous on aphids. *Ethology Ecology & Evolution* **10**, 361–368.

Agarwala, B. K. & Choudhuri, M. S. (1995) Use of alternative foods in the rearing of aphidophagous ladybeetle, *Menochilus sexmaculatus* Fabr. *Entomon* **20**, 19–23.

Agarwala, B. K., Das, S. & Senchowdhuri, M. (1988) Biology and food relations of *Micraspis discolor* (F.) an aphidophagous coccinellid in India. *Journal of Aphidology* **2**, 7–17.

Agarwala, B. K. & Dixon, A. F. G. (1992) Laboratory study of cannibalism and inter-specific predation in ladybirds. *Ecological Entomology* **17**, 303–309.

Agarwala, B. K. & Dixon, A. F. G. (1993) Why do ladybirds lay eggs in clusters? *Functional Ecology* **7**, 541–548.

Aguilera, A. (1995) Contribution to the knowledge of *Coccinella eryngii* (Mulsant) (Coleoptera: Coccinellidae) in Chile. *Acta Entomologica Chileana* **19**, 99–104.

Aguilera, A. & Diaz, G. (1983) Observaciones sobre la biología de *Neda patula* (Erichson) (Coleoptera: Coccinellidae) en Arica, Chile. *Indesia (Chile)* **7**, 25–38.

Ahmad, R. (1970) Studies in West Pakistan on the biology of one nitidulid species and two coccinellid species (Coleoptera) that attack scale insects (Hom., Coccoidea). *Bulletin of Entomological Research* **60**, 5–16.

Ahmad, R. & Ghani, M. A. (1966) Biology of *Chilocorus infernalis* Mulsant (Coleoptera Coccinellidae). *Technical Bulletin of the Commonwealth Institute of Biological Control* **7**, 101–106.

Al Abassi, S., Birkett, M. A., Pettersson, J., Pickett, J. A. & Woodcock, C. M. (1998) Ladybird beetle odour identified and found to be responsible for attraction between adults. *Cellular and Molecular Life Sciences* **54**, 876–879.

Alexander, R. D. (1974) The evolution of social behavior. *Annual Review of Ecology and Systematics* **5**, 325–383.

Alexander, R. D., Hoogland, J. L., Howard, R. D., Noonan, K. M. & Sherman, P. W. (1979) Sexual dimorphism and breeding systems in pinnipeds, ungulates, primates, an humans. In *Evolutionary Biology and Human Social Behaviour: An Anthropological Perspective*, ed. N. A. Chaguon & W. Irons, pp. 402–405. North Scituate, MA: Duxbury Press.

Ali, M. A. M. (1979) *Ecological and Physiological Studies on the Alfalfa Ladybird*. Budapest: Akademiai Kiado.

Alikan, M. A. & Yousuf, M. (1986) Temperature and food requirements of *Chilomenes sexmaculata* (Coleoptera : Coccinellidae). *Environmental Entomology* **15**, 800–802.

Allen, D. C., Knight, F. B. & Foltz, J. L. (1970) Invertebrate predators of the jack-pine budworm, *Choristoneura pinus*, in Michigan. *Annals of the Entomological Society of America* **63**, 59–64.

Anderson, J. M. E. (1980) Biology and distribution of *Scymnodes lividigaster* (Mulsant) and *Leptothea galbula* (Mulsant), Australian ladybirds (Coleoptera: Coccinellidae). *Proceedings of the Linnean Society of New South Wales* **105**, 1–15.

Arak, A. (1988) Sexual dimorphism in body size: a model and test. *Evolution* **42**, 820–825.

Arakaki, N. (1988) Egg protection with faeces in the ladybeetle, *Pseudoscymnus kurohime* (Miyatake) (Coleoptera: Coccinellidae). *Applied Entomology and Zoology* **23**, 495–497.

Arakaki, N. (1989) Alarm pheromone eliciting attack and escape responses in the sugar cane wooly aphid *Ceratovacuna lanigera* (Homoptera: Pemphigidae). *Journal of Ethology* **7**, 83–90.

Ayr, W. A. & Browne, L. M. (1977) The ladybug alakaloids including synthesis and biosynthesis. *Heterocycles* **7**, 685–707.

Babu, A. & Ananthakrishnan, T. N. (1993) Predatory efficiency, reproductive potential, and bioenergetics of *Coccinella transversalis* F. and *Menochilus sexmaculatus* F. (Coleoptera : Coccinellidae) in relation to aphid (Homoptera : Aphididae) prey quality. *Phytophaga* **5**, 121–133.

Bagal, S. R. & Trehan, K. N. (1945) Life-history and bionomics of two predaceous and one mycophagous species of Coccinellidae. *Journal of the Bombay Natural History Society* **45**, 566–575.

Bain, J., Singh, P., Ashby, M. D. & van Boven, R. J. (1994) Laboratory rearing of the predatory coccinellid *Cleobora mellyi* (Col.: Coccinellidae) for biological control of *Paropsis charybdis* (Col.: Chrysomelidae) in New Zealand. *Entomophaga* **29**, 237–244.

Balduf, W. V. (1935) *The Bionomics of Entomophagous Coleoptera.* St Louis, MO: Swift Company Inc.

Banks, C. J. (1954) The searching behaviour of coccinellid larvae. *Animal Behaviour* **2**, 37–38.

Banks, C. J. (1957) The behaviour of individual coccinellid larvae on plants. *Animal Behaviour* **5**, 12–24.

Barbier, R., Le Lannic, J. & Brun, J. (1996) Récepteurs sensoriels des palpes maxillaires de Coccinellidae adultes aphidiphages, coccidiphages et phytophages. *Bulletin de la Société zoologique de France* **121**, 255–268.

Barczak, T., Kaczorowski, G. & Burmistrzak, M. (1996) Coccinellid beetles (Coccinellidae, Col.) associated with population of *Aphis fabae* Scop. – complex (Aphididae, Hom.) on spindle bush. Preliminary results. In *Aphids and Other Homopterous Insects*, vol. 5 ed. J. Narkiewicz-Jodko, E. Cichocka, B. Nawrocka & W. Goszczynski, pp. 15–22. Skierniewice: Polish Academy of Science Vth Division – Agricultural and Forest Sciences.

Barron, A. & Wilson, K. (1998) Overwintering survival in the seven spot ladybird, *Coccinella septempunctata* (Coleoptera: Coccinellidae). *European Journal of Entomology* **95**, 639–642.

Bathon, H. & Pietrzik, J. (1986) Zur Nahrungsaufnahme des Bogen-Marienkafers, *Clitostethus arcuatus* (Rossi) (Col., Coccinellidae), einem Vertilger der Kohlmottenlaus, *Aleurodes proletella* Linne (Hom., Aleurodidae). *Journal of Applied Entomology* **102**, 321–326.

Baumgaertner, J. U., Gutierrez, A. P. & Summers, C. G. (1981) The influence of aphid prey consumption on searching behavior, weight increase, developmental time, and mortality of *Chrysopa carnea* (Neuroptera: Chrysopidae) and *Hippodamia convergens* (Coleoptera: Coccinellidae) larvae. *Canadian Entomologist* **114**, 1007–1014.

Baumgaertner, J. U., Bieri, M. & Delucchi, V. (1987) Growth and development of immature life stages of *Propylea 14-punctata* L. and *Coccinella 7-punctata* L. [Col.: Coccinellidae] simulated by the metabolic pool model. *Entomophaga* **32**, 415–423.

Beddington, J. R., Free, C. A. & Lawton, J. H. (1978) Modelling biological control: on the characteristics of successful natural enemies. *Nature* **273**, 513–518.

Bedford, E. C. G. (1949) Report of the plant pathologist. *Report of the Department of Agriculture, Bermuda*, pp. 11–19.

Begon, M., Harper, J. L. & Townsend, C. R. (1996) *Ecology. Individuals, Populations and Communities*. Oxford: Blackwell Scientific Publications.

Benham, B. R. & Muggleton, J. (1970) Studies on the ecology of *Coccinella undecimpunctata* Linn. (Col. Coccinellidae). *Entomologist* **103**, 153–170.

Benrey, B. & Denno, R. F. (1997) The slow-growth-high-mortality hypothesis: a test using the cabbage butterfly. *Ecology* **78**, 987–999.

Benrey, B. & Lamp, W. O. (1994) Biological control in the management of planthopper populations. In *Planthoppers: Their Ecology and Management*, ed. R. F. Denno & T. J. Perfect, pp. 519–550. New York: Chapman & Hall.

Berti, N., Boulard, M. & Duverger, C. (1983) Fourmis et Coccinelles: Revue bibliographique et observations nouvelles. *Bulletin de la Société entomologique de France* **88**, 271–274.

Blackburn, T. M. & Gaston, K. J. (1994) Animal body size distributions: patterns, mechanisms and implications. *Trends in Ecology and Evolution* **9**, 471–474.

Blackburn, T. M. & Gaston, K. J. (1996) On being the right size: different definitions of 'right'. *Oikos* **75**, 551–557.

Blackman, R. L. (1965) Studies on specificity in Coccinellidae. *Annals of Applied Biology* **56**, 336–338.

Blackman, R. L. (1967) The effects of different aphid foods on *Adalia bipunctata* L. and *Coccinella 7-punctata* L. *Annals of Applied Biology* **59**, 207–219.

Blattný, C. (1925) Het voorspellen van het masaal optredenvan schdelike insekten. *Tijdschrift Plantenziekten* **31**, 139–144.

Bodenheimer, F. S. (1932) *Icerya purchasi* Mask. und *Novius cardinalis* Muls. Eine bevölkerungswissenschaftliche Studie über die Grundlagen der biologischen Bekämpfung. *Zeitschrift für angewandte Entomologie* **19**, 514–543.

Bodenheimer, F. S. (1943) Studies on the life-history and ecology of Coccinellidae: 1. The life-history of *Coccinella septempunctata* L. in four different zoogeographical regions. *Bulletin Society Fouad 1er Entomology* **27**, 1–28.

Bodenheimer, F. S. (1951) *Citrus Entomology in the Middle East with Special References to Egypt, Iran, Irak, Palestine, Syria, Turkey*. Jerusalem: Dr W. Junk.

Bogdanova, N. L. (1956) *Hyperaspis campestris* Herbst (Coleoptera, Coccinellidae) as destroyer of *Chloropulvinaria floccifera* Westw. (Homoptera, Coccoidea). *Revue d' Entomologie de l'URSS* **35**, 311–323.

Boldyrev, M. J. & Wilde, W. H. A. (1969) Food seeking and survival in predaceous coccinellid larvae. *Canadian Entomologist* **101**, 1218–1222.

Bombosch, S. (1963) Untersuchungen zur Vermehrung von *Aphis fabae* Scop. in Samenrübenbeständen unter besondere Berücksichtigung der Schwebfliegen (Diptera: Syrphidae). *Zeitschrift für angewandte Entomologie* **52**, 105–141.

Bonnemaison, L. (1966) Integrated control of several aphids. In *Ecology of Aphidophagous Insects*, ed. I. Hodek, pp. 329–330. Prague: Academia.

Bonner, J. T. (1965) *Size and Cycle*. Princeton, NJ: Princeton University Press.

Booth, R. G. (1997) A review of the species of *Calvia* (Coleoptera: Coccinellidae) from the Indian subcontinent, with descriptions of two new species. *Journal of Natural History* **31**, 917–934.

Booth, R. G., Cross, A. E., Fowler, S. V. & Shaw, R. H. (1995) The biology and taxonomy of *Hyperaspis pantherina* (Coleoptera: Coccinellidae) and the classical biological control of its prey *Orthezia insignis* (Homoptera: Ortheziidae). *Bulletin of Entomological Research* **85**, 307–314.

Bose, K. C. & Ray, S. K. (1967) Aphid–predator balance – II. A comparative study on the consumption of aphids by the common predator, *Cheilomenes sexmaculata* Fabr. (Coleoptera: Coccinellidae). *Indian Journal of Science and Industry* **1**, 56–59.

Boving, A. G. & Craighead, F. C. (1930) An illustrated synopsis of the principal larval forms of the order Coleoptera. *Entomologica Americana* **11**, 1–351.

Brafield, A. E. & Llewellyn, M. J. (1982) *Animal Energetics*. Glasgow: Blackie.

Brakefield, P. M. (1985) Polymorphic Müllerian mimicry and interactions with thermal melanism in ladybirds and a soldier beetle: a hypothesis. *Biological Journal of the Linnean Society* **26**, 243–267.

Brakefield, P. M. & Willmer, P. G. (1985) The basis of thermal melanism in the ladybird *Adalia bipunctata*. Differences in reflectance and thermal properties between morphs. *Heredity* **54**, 9–14.

Brettell, J. H. (1964) Biology of *Diloponis inconspicuus* Pope (Coleoptera : Coccinellidae), a predator of citrus red scale, with notes on the feeding beahiour of other scale predators. *Journal of the Entomological Society of South Africa* **27**, 17–28.

Bristow, C. M. (1988) What makes a predator specialize? *Trends in Ecology and Evolution* **3**, 1–2.

Brown, H. D. (1972) On the biology of *Lioadalia flavomaculata* (De G.) (Col. , Coccinellidae), a predator of the wheat aphid (*Schizaphis graminum* (Rond.)) in South Africa. *Bulletin of Entomological Research* **61**, 673–679.

Brown, J. H., Marquet, P. A. & Taper, M. L. (1993). Evolution of body size: consequences of an energetic definition of fitness. *American Naturalist* **142**, 573–584.

Bugg, R. L. & Dutcher, J. D. (1993). *Sesbania exaltata* (Rafinesque-Schmaltz) Cory (Fabaceae) as a warm-season cover crop in pecan orchards: effects on aphidophagous Coccinellidae and pecan aphids. *Biological Agriculture and Horticulture* **9**, 215–229.

Bugg, R. L., Dutcher, J. D. & McNeill, P. J. (1991) Cool-season cover crops in the pecan orchard understory: effects on Coccinellidae (Coleoptera) and pecan aphids (Homoptera: Aphididae). *Biological Control* **1**, 8–15.

Buntin, L. A. & Tamaki, G. (1980) Bionomics of *Scymnus marginicollis* (Coleoptera: Coccinellidae). *Canadian Entomologist* **112**, 675–680.

Butler, G. D. (1982) Development time of *Coccinella septempunctata* in relation to constant temperatures [Col.: Coccinellidae]. *Entomophaga* **27**, 349–353.

Butler, G. D. & Dickerson, W. A. (1972) Life cycle of the convergent lady beetle in relation to temperature. *Journal of Economic Entomology* **65**, 1508–1509.

Calder, W. A. (1984). *Size, Function and Life History*. Cambridge, MA: Harvard University Press.

Calunga-Garcia, M. & Gage, S. H. (1999) Arrival, establishment, and habitat use of the multicolored asian lady beetle (Coleoptera: Coccinellidae) in a Michigan landscape. *Environmental Entomology* **27**, 1574–1580.

Camargo, F. C. (1937) Notas taxonomicas e biologicas sobra algunos Coccinellideos do genero *Neocalvia* Crotch, predatores de las larvas do genero *Psyllobora* Chevrolat (Col. Coccinellidae). *Revista Entomologia Rio de Janeiro* **7**, 362–377.

Campbell, A., Frazer, B. D., Gilbert, N., Gutierrez, A. P. & Mackauer, M. (1974) The temperature requirements of some aphids and their parasites. *Journal of Applied Ecology* **11**, 431–438.

Cambell, R. K., Farris, T. N., Perring, T. M., Leonard, M. E., Cartwright, B. O. & Eikenbary, R. D. (1980) Biological observations of *Menochilus sexmaculatus*, reared on *Schizaphis graminum*. *Annals of the Entomological Society of America* **73**, 153–157.

Carness, E. K. (1912a) Collecting ladybirds (Coccinellidae) by the ton. *Monthly Bulletin California Commn Horticulture* **1**, 71–81.

Carness, E. K. (1912b) An explanation of the hibernating habits of *Hippodamia convergens*. *Monthly Bulletin California Commn Horticulture* **1**, 177–188.

Carter, M. C. (1982) The foraging strategy of *Coccinella septempunctata* (L.). PhD thesis, University of East Anglia, U.K.

Carter, M. C. & Dixon, A. F. G. (1982) Habitat quality and the foraging behaviour of coccinellid larvae. *Journal of Animal Ecology* **51**, 865–878.

Carter, M. C. & Dixon, A. F. G. (1984) Honeydew: an arrestant stimulus for coccinellids. *Ecological Entomology* **9**, 383–387.

Carter, M. C., Sutherland, D. & Dixon, A. F. G. (1984) Plant structure and the searching efficiency of coccinellid larvae. *Oecologia* **63**, 394–397.

Carter, N., Dixon, A. F. G. & Rabbinge, R. (1982) *Cereal Aphid Populations: Biology, Simulation and Prediction.* Wageningen: Pudoc.

Chakrabarti, S., Debnath, N. & Ghosh, D. (1995) Bioecology of *Harmonia eucharis* (Mulsant) (Coleoptera: Coccinellidae). An aphidophagous predator in Western Himalayas. *Entomon* **20**, 191–196.

Chakrabarti, S., Ghosh, D. & Debnath, N. (1988) Developmental rate and larval voracity in *Harmonia (Leis) dimidiata* (Col., Coccinellidae), a predator of *Eriosoma lanigerum* (Hom., Aphididae) in western Himalaya. *Acta Entomologica Bohemoslovaca* **85**, 335–339.

Chang, G. C. (1996) Comparison of single versus multiple species of generalist predators for biological control. *Environmental Entomology* **25**, 207–212.

Charnov, E. L. (1976a) Optimal foraging: the marginal value theorem. *Theoretical Population Biology* **9**, 126–136.

Charnov, E. L. (1976b) Optimal foraging: attack strategy of a mantid. *American Naturalist* **110**, 141–151.

Charnov, E. L., Orians, G. H. & Hyatt, K. (1976) Ecological implications of resource depression. *American Naturalist* **110**, 247–259.

Chazeau, J. (1974) Évaluation de l'action prédatrice de *Stethorus madecassus* [Coleoptère Coccinellidae] sur *Tetranychidae neocaledonicus* [Acarien Tetranychidae]. *Entomophaga* **19**, 183–193.

Chazeau, J. (1981). Données sur la biologie de *Coelophora quadrivittata* (Col. : Coccinellidae) en Nouvelle Calédonie. *Entomophaga* **26**, 301–302.

Chazeau, J., Bouye, E. & Bonnet De Larbogne, L. (1991) Cycle de développement et table de vie d' *Olla v-nigrum* [Col. : Coccinellidae] ennemi naturel d' *Heteropsylla cubana* [Hom. : Psyllidae] introduit en Nouvelle Calédonie. *Entomophaga* **36**, 275–285.

Cheah, C. A. S.-J. & McClure, M. S. (1999) Life history and development of *Pseudoscymnus tsugae* (Coleoptera: Coccinellidae), a new predator of the hemlock woolly adelgid (Homoptera: Adelgidae). *Environmental Entomology* **27**, 1531–1536.

Cisneros, J. J. & Rosenheim, J. A. (1997) Ontogenetic change of prey preference in a generalist predator, *Zeleus renardii*, and its influence on the intensity of predator–predator interactions. *Ecological Entomology* **22**, 399–407.

Cividanes, F. J. & Gutierrez, A. P. (1996) Modelling the age-specific per capita growth and reproduction of *Rhizobius lophanthae* (Col.: Coccinellidae). *Entomophaga* **41**, 257–266.

Clancy, K. M. & Price, P. W. (1987) Rapid herbivore growth enhances enemy attack: sublethal defenses remain a paradox. *Ecology* **68**, 736–738.

Clark, T. L. & Messina, F. J. (1998) Plant architecture and the foraging success of ladybird beetles attacking the Russian wheat aphid. *Entomologia Experimentalis et Applicata* **86**, 153–161.

Clausen, C. P. (1940) *Entomophagous Insects*. New York: McGraw-Hill.

Cochereau, P. (1969) Contrôle biologique d'*Aspidiotus destructor* Signoret (Homoptera, Diaspinae) dans l'île Vaté (Nouvelles Hébrides) au moyen de *Rhizobius pulchellus* Montrouzier (Coleoptera, Coccinellidae). *Cahiers de ORSTOM séries Biologique* **8**, 57–100.

Colburn, R. & Asquith, D. (1970) A cage used to study the finding of a host by the ladybeetle, *Stethorus punctum*. *Journal of Economic Entomology* **63**, 1376–7.

Colfer, R. G. & Rosenheim, J. A. (1995) Intraguild predation by coccinellid beetles on an aphid parasitoid, *Lysiphlebus testaceipes*. In *1995 Proceedings Beltwide Cotton Conferences*, vol. 2, 1033–1036.

Collins, M. D., Ward, S. A. & Dixon, A. F. G. (1981) Handling time and the functional response of *Aphelinus thomsoni*, a predator and parasite of the aphid *Drepanosiphum platanoidis*. *Journal of Animal Ecology* **50**, 479–487.

Collyer, E. (1953). Biology of some predatory insects and mites associated with the fruit tree red spider mite (*Metatetranychus ulmi* (Koch)) in South-Eastern England II. Some important predators of the mite. *Journal of Horticultural Science* **28**, 85–97.

Commins, H. N. & Hassell, M. P. (1979) The dynamics of optimally foraging predators and parasitoids. *Journal of Animal Ecology* **48**, 335–351.

Cook, R. M. & Cockrell, B. J. (1978) Predator ingestion rate and its bearing on feeding time and the theory of optimal diets. *Journal of Animal Ecology* **47**, 529–549.

Cook, R. M. & Hubbard, S. F. (1977) Adaptive searching strategies in insect parasites. *Journal of Animal Ecology* **46**, 115–125.

Cornell, H. (1976) Search strategies and the adaptive significance of switching in some general predators. *American Naturalist* **111**, 317–320.

Correjo, N. S. G., Bellotti, A. C. & Gonzalez, R. O. (1991) Evaluación de algunos factores determinantes de la eficiencia de *Cleothera notata* (Col: Coccinellidae) como depredador del piojo harinoso de la yuca *Phenacoccus herreni* (Hom: Pseudococcidae). *Revista Colombiana de Entomología* **17**, 21–27.

Cottrell, T. E. & Yeargan, K. V. (1999) Influence of a native weed, *Acalypha ostryaefolia* (Euphorbiaceae), on *Coleomegilla maculata* (Coleoptera: Coccinellidae) population density, predation, and cannibalism on sweet corn. *Environmental Entomology* **27**, 1375–1385.

Crawley, M. J. (ed.) (1992) *Natural Enemies: The Population Biology of Predators, Parasites and Diseases*. Oxford: Blackwell Scientific Publications.

Crotch, G. R. (1874) *A Revision of the Coleopterous Family Coccinellidae*. London: Janson.

Crowson, R. A. (1981) *The Biology of Coleoptera*. London: Academic Press.

Daloze, D., Braekman, J.-C. & Pasteels, J. M. (1995) Ladybird defence alkaloids: structural, chemotaxonomic and biosynthetic aspects (Col.: Coccinellidae). *Chemoecology* **5/6**, 173–183.

Danks, H. V. (1987) *Insect Dormancy: An Ecological Perspective*. Ottawa: Biological Survey of Canada (Terrestrial Arthropods), National Museum of Natural Sciences.

Davidson, D. W. (1977). Species diversity and community organization in desert seed-eating ants. *Ecology* **58**, 711–724.

Davidson, W. M. (1923). Biology of *Scymnus nubes* Casey (Coleoptera, Coccinellidae). *Transactions of the American Entomological Society* **49**, 155–163.

DeBach, P. (1958) Selective breeding to improve adaptations of parasitic insects. In *Proceedings of the 10th International Congress of Entomology*, vol. 4, p. 759.

DeBach, P. (1974) *Biological Control by Natural Enemies*. Cambridge: Cambridge University Press.

Delucchi, V. (1954). *Pullus impexus* (Muls.) (Coleoptera: Coccinellidae), a predator of

Adelges picea (Ratz.) (Hemiptera, Adelgidae) with notes on its parasites. *Bulletin of Entomological Research* **45**, 243–278.

Dimetry, N. Z. (1974) The consequences of egg cannibalism in *Adalia bipunctata* (Coleoptera: Coccinellidae). *Entomophaga* **19**, 445–451.

Dimetry, N. Z. (1976) The role of predator and prey density as factors affecting behavioural and biological aspects of *Adalia bipunctata* (L.) larvae. *Zeitschrift für angewandte Entomologie* **81**, 386–392.

Dimetry, N. Z. & Mansour, M. H. (1976) The choice of oviposition sites by ladybird beetle *Adalia bipunctata* (L.). *Experientia* **32**, 181–182.

Ding-Xin, Z. & Zhong-Wen, W. (1987) Influence of temperature on the development of the coccinellid beetle, *Scymnus hoffmanni* Weise. *Acta Entomologica Sinica* **30**, 47–54.

Dixon, A. F. G. (1958) The escape responses shown by certain aphids to the presence of the coccinellid *Adalia decempunctata* (L.). *Transactions of the Royal Entomological Society of London* **110**, 319–334.

Dixon, A. F. G. (1959) An experimental study of the searching behaviour of the predatory coccinellid beetle *Adalia decempunctata* (L.). *Journal of Animal Ecology* **28**, 259–281.

Dixon, A. F. G. (1987) Parthenogenetic reproduction and the rate of increase in aphids. In *Aphids, their Biology, Natural Enemies and Control*, ed. A. K. Minks & P. Harrewijn, pp. 269–287. Amsterdam: Elsevier.

Dixon, A. F. G. (1997) Patch quality and fitness in predatory ladybirds. *Ecological Studies* **130**, 205–223.

Dixon, A. F. G. (1998) *Aphid Ecology*. London: Chapman & Hall.

Dixon, A. F. G. & Agarwala, B. K. (1999) Ladybird-induced life history changes in aphids. *Proceedings of the Royal Society London* **B 266**, 1–5.

Dixon, A. F. G. & Guo, Y. (1993) Egg and cluster size in ladybird beetles (Coleoptera: Coccinellidae): The direct and indirect effects of aphid abundance. *European Journal of Entomology* **90**, 457–463.

Dixon, A. F. G & Hemptinne, J.-L. (2000) Body size distribution in predatory ladybird beetles reflects that of their prey. *Ecology*

Dixon, A. F. G., Hemptinne, J.-L. & Kindlmann, P. (1997) Effectiveness of ladybirds as biological control agents: patterns and processes. *Entomophaga* **42**, 71–83.

Dixon, A. F. G. & Kindlmann, P. (1998a) Population dynamics of aphids. In *Insect Populations*, ed. J. P. Dempster & I. F. G. McLean, pp. 207–230. Dordrecht: Kluwer Academic Publishers.

Dixon, A. F. G. & Kindlmann, P. (1998b) Generation time ratio and the effectiveness of ladybirds as classical biological control agents. In *Proceedings of the Australasian Applied Entomological Research Conference*, vol. 1, ed. M. P. Zalucki, R. A. I. Drew & G. G. White, pp. 314–320.

Dixon, A. F. G. & Kindlmann, P. (1999) Cost of flight apparatus and optimum body size of aphid migrants. *Ecology*

Dixon, A. F. G., Kindlmann, P. & Jarosik, V. (1995) Body size distribution in aphids: relative surface area of specific plant structures. *Ecological Entomology* **20**, 111–117.

Dixon, A. F. G. & Kundu, R. (1997) Trade-off between reproduction and length of adult life in males and mating females in aphids. *European Journal of Entomology* **94**, 105–109.

Dixon, A. F. G. & McKay, S. (1970) Aggregation in the sycamore aphid *Drepanosiphum platanoides* (Schr.) (Hemiptera: Aphididae) and its relevance to the regulation of population growth. *Journal of Animal Ecology* **39**, 439–454.

Dixon, A. F. G. & Russel, R. J. (1972). The effectiveness of *Anthocoris nemorum* and *A. confusus* (Hemiptera: Anthocoridae) as predators of the sycamore aphid,

Drepanosiphum platanoides II. Searching behaviour and the incidence of predation in the field. *Entomologia Experimentalis et Applicata* **15**, 35–50.

Dixon, A. F. G. & Stewart, L. A. (1991) Size and foraging in ladybird beetles. In *Behaviour and Impact of Aphidophaga*, ed. L. Polgár, R. J. Chambers, A. F. G. Dixon & I. Hodek, pp. 123–132. The Hague: SPB Academic Publishing.

Dong, Q. & Polis, G. A. (1992) The dynamics of cannibalistic populations: a foraging perspective. In *Cannibalism Ecology and Evolution among Diverse Taxa*, eds M. A. Elgar & B. J. Crespi, pp. 13–37. Oxford: Oxford University Press.

Doumbia, M., Hemptinne, J.-L. & Dixon, A. F. G. (1998) Assessment of patch quality by ladybirds : rôle of larval tracks. *Oecologia* **113**, 197–202.

Doutt, R. L. (1964) The historical development of biological control. In *Biological Control of Insect Pests and Weeds*, ed. P. DeBach, pp. 21–42. London: Chapman & Hall.

Van Driesche R. G. & Bellows, T. S. (1996) *Biological Control*. New York: Chapman & Hall.

Dreistadt, S. H. & Flint, M. L. (1996) Melon aphid (Homoptera: Aphididae) control by inundative convergent lady beetle (Coleoptera: Coccinellidae) release on chrysanthemum. *Environmental Entomology* **25**, 688–697.

Duelli, P. (1981) Is larval cannibalism in lacewings adaptive? (Neuroptera: Chrysopidae). *Researches in Population Ecology* **23**, 193–209.

Eastop, V. F. (1978). Diversity of the Sternorrhyncha within major climatic zones. In *Diversity of Insect Faunas*, ed. L. A. Mound & N. Waloff, pp. 71–87. Symposium of the Royal Entomological Society London no. 9.

Ehler, L. E. (1990) Introduction strategies in biological control of insects. In *Critical Issues in Biological Control*, ed, M. Mackauer, L. E. Ehler & J. Roland, pp. 111–134. Andover: Intercept.

Eisner, T., Ziegler, R. , McCormick, J. L., Eisner, M., Hoebeke, E. R. & Meinwald, J. (1994) Defensive use of an acquired substance (carminic acid) by predaceous insect larvae. *Experientia* **50**, 610–615.

Elgar, M. A. & Crespi, B. J. (1992) Ecology and evolution of cannibalism. In *Cannibalism Ecology and Evolution in Diverse Taxa*, ed. M. A. Elgar & B. J. Crespi, pp. 1–12. Oxford: Oxford University Press.

Ellingsen, I.-J. (1960) Fecundity, aphid consumption and survival of the aphid predator *Adalia bipunctata* L. (Col., Coccinellidae). *Norwegian Journal of Entomology* **16**, 91–95.

El Hag, E. A. & Zaitoon, A. A. (1996) Biological parameters for four coccinellid species in Central Saudi Arabia. *Biological Control* **7**, 316–319.

El-Hariri, G. (1966) Studies of the physiology of hibernating Coccinellidae (Coleoptera): changes in the metabolic reserves and gonads. *Proceedings of the Royal Entomological Society of London A* **41**, 133–144.

Elliott, N., Kieckhefer, R. & Kauffman, W. (1996) Effects of an invading coccinellid on native coccinellids in an agricultural landscape. *Oecologia* **105**, 537–544.

Elliott, H. J. & de Little, D. W. (1980) Laboratory studies on predation of *Chrysophtharta bimaculata* (Olivier) (Coleoptera: Chrysomelidae) eggs by the coccinellids *Cleobora mellyi* (Mulsant) and *Harmonia conformis* (Boisduval). *General Applied Entomology* **12**, 33–36.

Elton, C. S. (1927). *Animal Ecology*. London: Sidgwick & Jackson.

van Emden, H. F. (1966) The effectiveness of aphidophagous insects in reducing aphid populations. In *Ecology of Aphidophagous Insects*, ed. I. Hodek, pp. 227–235. Prague: Academia.

Emrich, B. H. (1991) Erworbene Toxizität bei der Lupinenblattlaus *Macrosiphum albifrons* und ihr Einfluss auf die aphidophagen Prädatoren *Coccinella septempunctata, Episyrphus balteatus* und *Chrysoperla carnea. Zeitschrift für Pflanzenkrankheiten und Pflanzenschutz* **98**, 398–404.

Ericksen, C., Samways, M. J. & Hattingh, V. (1991) Reaction of the ladybird *Chilocorus nigritus* (F.) (Col., Coccinellidae) to a doomed food resource. *Journal of Applied Entomology* **112**, 493–498.

Evans, E. W. (1991) Intra- versus interspecific interactions of ladybeetles (Coleoptera: Coccinellidae) attacking aphids. *Oecologia* **87**, 401–408.

Evans, E. W. & Dixon, A. F. G. (1986) Cues for oviposition by ladybird beetles (Coccinellidae): response to aphids. *Journal of Animal Ecology* **55**, 1027–1034.

Fabres, G. & Kiyindou, A. (1985) Comparison du potentiel biotique de deux coccinelles (*Exochomus flaviventris* et *Hyperaspis senegalensis hottentotta* Col. Coccinellidae) prédatrices de *Phenacoccus manihoti* (Hom. Pseudococcidae) au Congo. *Acta Oecologia* **6**, 339–348.

Fairbairn, D. J. (1990) Factors influencing sexual dimorphism in temperate waterstriders. *American Naturalist* **136**, 61–86.

Ferguson, K. I. & Stiling, P. (1996) Non-additive effects of multiple natural enemies on aphid populations. *Oecologia* **108**, 375–379.

Ferran, A., Cruz de Boelpaepe, M. O., Buscarlet, L. A., Larroque, M. M. & Schanderl, H. (1984) Les relations trophiques entre les larves de la coccinelle *Semiadalia undecimnotata* Schn. et le puceron *Myzus persicae* Sulz.: généralisation à d'autres couples 'proie–prédateur' et influence des conditions d'élevage de l'auxiliare. *Acta Oecologia / Oecologia Applicata* **5**, 85–97.

Ferran, A. & Dixon, A. F. G. (1993) Foraging behaviour of ladybird larvae (Coleoptera: Coccinellidae). *European Journal of Entomology* **90**, 383–402.

Ferran, A., Gambier, J., Parent, S., Legendre, K., Tourniaire, R. & Giuge, L. (1997a) The effect of rearing the ladybird *Harmonia axyridis* on *Ephestia kuehniella* eggs on the response of its larvae to aphid tracks. *Journal of Insect Behavior* **10**, 129–143.

Ferran, A., Giuge, L., Tourniaire, R., Gambier, J. & Fournier, D. (1998) An artificial non-flying mutation to improve the efficiency of the ladybird *Harmonia axyridis* in biological control of aphids. *Biocontrol* **43**, 53–64.

Ferran, A. & Larroque, M. M. (1977) Sur une possibilité d'estimer l'action prédatrice des larves de la Coccinelle Aphidophage *Semiadalia 11-notata* Schn. (Col. Coccinellidae) grâce à la connaissance de leur évolution pondérale. *Annales de Zoologie et Écologie animale* **9**, 693–708.

Ferran, A. & Larroque, M. M. (1979) Influences des facteurs abiotiques sur la physiologie alimentaire des larves de la coccinelle aphidiphage *Semiadalia undecimnotata* Schn. (Col., Coccinellidae) 1. Action de la température. *Entomophaga* **24**, 403–410.

Ferran, A., Nikham, H., Kabiri, F., Picart, J. L., De Herce, C., Brun, J., Iperti, G. & Lapchin, L. (1997b) The use of *Harmonia axyridis* larvae (Coleoptera, Coccinellidae) against *Macrosiphum rosae* (Hemiptera, Aphididae) on rose bushes. *European Journal of Entomology* **93**, 59–67.

Fleschner, C.A. (1950) Studies on searching capacity of the larvae of three predators of the citrus red mite. *Hilgardia* **20**, 233–265.

Flint, M. L., Dreistadt, S. H., Rentner, J. & Parella, M. P. (1995) Lady beetle release controls aphids on potted plants. *California Agriculture* **49**, 5–8.

de Fluiter, H. J. (1939) Beitrag zur Kenntnis der Biologie und Ökologie einiger Coccinelliden von Java. In *7th International Kongress Entomologie, Berlin (1939)*, pp. 1106–1119.

Formusoh, E. S. & Wilde, G. E. (1993) Preference and development of two species of predatory coccinellid on the Russian wheat aphid and greenbug biotype E (Homoptera: Aphididae). *Journal of Agricultural Entomology* **10**, 65–70.

Fox, L. R. (1975) Cannibalism in natural populations. *Annual Review of Ecology and Systematics* **6**, 87–106.

Frank, J. H. (1998) How risky is biological control? Comment. *Ecology* **79**, 1829–34.

Frazer, B. D. (1988) Coccinellidae. In *Aphids: Their Biology, Natural Enemies and Control*, vol. 2B, ed. A. K. Minks & P. Harrewijn, pp.231–247. Amsterdam: Elsevier.

Frazer, B. D. & Gilbert, N. (1976) Coccinellids and aphids: a quantitative study of the impact of adult lady-birds (Coleoptera: Coccinellidae) preying on field populations of pea aphids (Homoptera: Aphididae). *Journal of the Entomological Society of British Columbia* **73**, 33–56.

Frazer, B. D., Gilbert, N., Ives, P. M. & Ranworth, D. A. (1981) Predator reproduction and overall predator–prey relationship. *Canadian Entomologist* **113**, 1015–1024.

Frazer, B. D. & Gill, B. (1981) Hunger, movement, and predation of *Coccinella californica* on pea aphids in the laboratory and in the field. *Canadian Entomologist* **113**, 1025–1033.

Frazer, B. D. & McGregor, R. R. (1994) Searching behaviour of adult female Coccinellidae (Coleoptera) on stem and leaf models. *Canadian Entomologist* **126**, 389–399.

Freier, B., Möwes, M. & Trilitsch, H. (1998) Beneficial thresholds for *Coccinella 7–punctata* L. (Col., Coccinellidae) as a predator of cereal aphids in winter wheat – results of population investigations and computer simulations. *Journal of Applied Entomology* **122**, 213–217.

Frogatt, W. W. (1902) Australian ladybird beetles. *Agricultural Gazette of New South Wales* **13**, 895–911.

Fürsch, H. (1967). Familie: Coccinellidae (Marienkäfer). In *Die Käfer Mitteleuropas. Band 7, Clavicornia*, ed. H. Freude, K. W. Harde, & G. A. Lohse, pp. 227–278. Krefeld: Goecke & Evers.

Gage, J. H. (1920) Larvae of Coccinellidae. *University of Illinois Biological Monographs* **6**, 232–294.

Gautam, R. D. (1990) Effect of substrata and age of *Menochilus sexmaculatus* (Fabr.) (Coleoptera: Coccinellidae) on its oviposition. *Journal of Biological Control* **4**, 8–10.

Gauthier, C. (1996) Un exemple d'adaptation chez les Coccinellidae: la protection chimique des oeufs. Diplôme d'Ingénieur Agronome de la Faculté universitaire des Sciences agronomiques, Gembloux, Belgium.

Gawande, R. B. (1966) Effect of constant and alternating temperatures on feeding and development of *Chilomenes sexmaculata* FB. In *Ecology of Aphidophagous Insects*, ed. I. Hodek, pp.63–67. Prague: Academia.

Geyer, J. W. C. (1947a) A study of the biology and ecology *Exochomus flavipes* Thunb. (Coccinellidae, Coleoptera) Part I. *Journal of the Entomological Society of South Africa* **9**, 219–234.

Geyer, J. W. C. (1947b) A study of the biology and ecology of *Exochomus flavipes* Thunb. (Coccinellidae, Coleoptera) Part II. *Journal of the Entomological Society of South Africa* **10**, 64–109.

Ghani, M. A. & Ahmad, R. (1966) Biology of *Pharoscymnus flexibilis* Muls. (Col. Coccinellidae). *Technical Bulletin of the Commonwealth Institute of Biological Control* **31**, 107–111.

Ghiselin, M. T. (1974) *The Economy of Nature and the Evolution of Sex*. Berkeley, CA: University of California Press.

Ghorpade, K. D. (1979) *Ballia eucharis* (Coleoptera : Coccinellidae) breeding on Cicadellidae (Homoptera) at Shillong. *Current Research* **8**, 113.

Gibson, R. L., Elliot, N. C. & Schaefer P. (1992) Life history and development of *Scymnus frontalis* (Fabricius) (Coleoptera: Coccinellidae) on four species of aphid. *Journal of the Kansas Entomological Society* **65**, 410–415.

Gilliam, J. (1990) Hunting by the hunted: optimal prey selection by foragers under predation hazard. In *Behavioral Mechanisms of Food Selection*, ed. R. Hughes, pp. 797–820. NATO ASI Series, Series A: Life Sciences, New York: Springer-Verlag.

Gilliam, J. & Fraser, D. (1987) Habitat selection under predation hazard: test of a model with foraging minnows. *Ecology* 68, 1856–1862.

Ginting, C. U., Djamin, A. & Sidauruk, L. (1992) Some biological aspects of *Megalocaria dilatata* Fab. (Col.: Coccinellidae) a potential coconut leaf aphid predator. *Oléagineux* 47, 569–572.

Glen, D.M. (1973). The food requirements of *Blepharidopterus angulatus* (Heteroptera: Miridae) as a predator of the lime aphid, *Eucallipterus tiliae. Entomologia Experimentalis et Applicata* 16, 255–267.

Godeau, F. & Hemptinne, J.-L. (2000) Adaptation of *Coccinella magnifica* to live with *Formica rufa. Insectes Sociaux*

Godfray, H. C. J. (1987) The evolution of clutch size in invertebrates. *Oxford Survey of Evolutionary Biology* 4, 117–154.

Godfray, H. C. J. (1994) *Parasitoids, Behavioral and Evolutionary Ecology.* Princeton, NJ: Princeton University Press.

Godfray, H. C. J. & Hassell, M. P. (1987) Natural enemies may be a cause of discrete generations in tropical insects. *Nature* 327, 144–147.

van Goethem, J. L. (1975). *Lieveheersbeestjetabel Coccinellidae van Belgie.* Brussels: Koninklijk Belgisch Instituut voor Natuurwetenschappen.

Gordon, R. D. (1976) The Scymnini (Coleoptera: Coccinellidae) of the United States and Canada: Key to genera and revision of *Scymnus, Nephus* and *Diomus. Bulletin of the Buffalo Society of Natural Sciences* 28, 1–362.

Gordon, R. D. (1985) The Coccinellidae (Coleoptera) of America north of Mexico. *Journal of the New York Entomological Society* 93, 1–912.

Gordon, R. D. & Vandenberg, N. (1993) Larval systematics of North American *Cycloneda* Crotch (Coleoptera: Coccinellidae). *Entomologica scandinavica* 24, 301–312.

Gordon, R. D. & Vandenberg, N. (1995) Larval systematics of North American *Coccinella* L. (Coleoptera : Coccinellidae). *Entomologica scandinavica* 26, 67–86.

Gould, S. J. (1988). Trends as changes in variance: a new slant on progress and directionality in evolution. *Journal of Paleontology* 62, 315–329.

Goureau, M. (1974) Systématique de la tribu des Scymnini (Coccinellidae). *Annales de Zoologie et Ecologie animale,* 223 pp.

Greathead, D. J. & Pope, R. D. (1977) Studies on the biology and taxonomy of some *Chilocorus* spp. (Coleoptera: Coccinellidae) preying on *Aulacaspis* spp. (Hemiptera: Diaspididae) in East Africa, with the description of a new species. *Bulletin of Entomological Research* 67, 259–270.

Grevstad, F. S. & Klepetka, B. W. (1992) The influence of plant architecture on the foraging efficiencies of a suite of ladybird beetles feeding on aphids. *Oecologia* 92, 399–404.

Grill, C. P. & Moore, A. J. (1998) Effects of a larval antipredator response and larval diet on adult phenotype in an aposematic ladybird beetle. *Oecologia* 114, 274–282.

Grill, C. P., Moore, A. J. & Brodie, E. D. (1997) The genetics of phenotypic plasticity in a colonizing population of the ladybird beetle, *Harmonia axyridis. Heredity* 78, 261–269.

Gurney, B. & Hussey, N. W. (1970) Evaluation of some coccinellid species for the biological control of aphids in protected cropping. *Annals of Applied Biology* 65, 451–458.

Gutierrez, A. P. (1996) *Applied Population Ecology: A Supply–Demand Approach.* New York: John Wiley.

Gutierrez, A. P., Baumgaertner, J. U. & Hagen, K. S. (1981) A conceptual model for growth, development, and reproduction in the ladybird beetle, *Hippodamia convergens* (Coleoptera: Coccinellidae). *Canadian Entomologist* 113, 21–33.

Gutierrez, A. P., Baumgaertner, J. U. & Summers, C. G. (1984) Multitrophic models of predator–prey energetics. *Canadian Entomologist* **116**, 923–963.

Gutierrez, J. & Chazeau, J. (1972) Cycles de développement et tables de vie de *Tetranychus neocaledonicus* André [Acariens : Tetranychidae] et d'un de ses principaux prédateurs à Madagascar *Stethorus madecassus* Chazeau [Coccinellidae]. *Entomophaga* **17**, 275–295.

Gutierrez, A. P. , Mills, N. J., Schreiber, S. J. & Ellis, C. K. (1994) A physiological based tritrophic perspective on bottom up/top down regulation of populations. *Ecology* **75**, 2227–2242.

Gutierrez, A. P., Neuenschwander, P., Schulthess, F., Herren, H. R., Baumgaertner, J. U., Wermelinger, B., Lohr, B. & Ellis, C. K. (1988) Analysis of biological control of cassava pests in Africa II. Cassava mealybug *Phenacoccus manihoti. Journal of Applied Ecology* **25**, 921–940.

Hafez, M. & El-Ziady, S. (1952) Studies on the biology of *Hyperaspis vimciguerrae* Capra, with a full description of the anatomy of the fourth stage larva. *Bulletin Society Fouad 1er Entomology* **36**, 211–246.

Hagen, K. S. (1962) Biology and ecology of predaceous Coccinellidae. *Annual Review of Entomology* **7**, 289–326.

Hagen , K. S. (1966) Suspected migratory flight behaviour of *Hippodamia convergens*. In *Ecology of Aphidophagous Insects*, ed. I. Hodek, pp. 135–136. Prague: Academia.

Hagen, K. S. (1974) The significance of predaceous *Coccinellidae* in biological and integrated control of insects. *Entomophaga* **7**, 25–44.

Hagen, K. S. & van den Bosch, R. (1968) Impact of pathogens, parasites, and predators on aphids. *Annual Review of Entomomogy* **13**, 325–384.

Hagen, K. S. & Sluss, R. R. (1966) Quantity of aphids required for reproduction by *Hippodamia* spp. in the laboratory. In *Ecology of Aphidophagous Insects*, ed. I. Hodek, pp. 47–59. Prague: Academia.

Hajek, A. E. & Dahlsten, D. L. (1987) Behavioural interactions between three birch aphid species and *Adalia bipunctata* larvae. *Entomologia Experimentalis et Applicata* **45**, 81–87.

Haldane, J. B. S. (1927). On being the right size. Reprinted in *On Being The Right Size and Other Essays*, ed. J. Maynard Smith, 1985. Oxford: Oxford University Press.

Hales, D. (1979) Population dynamics of *Harmonia conformis* (Boisd.) (Coleoptera: Coccinellidae) on acacia. *General Applied Entomology* **11**, 3–8.

Hall, S. J. & Raffaelli, D. G. (1993) Food webs: theory and reality. *Advances in Ecological Research* **24**, 187–239.

Hamilton, W. D. (1964) The genetical evolution of social behaviour 1. *Journal of Theoretical Biology* **7**, 1–16.

Harmon, J. P., Losey, J. E. & Ives, A. R. (1998) The role of vision and color in the close proximity foraging behavior of four coccinellid species. *Oecologia* **115**, 287–292.

Harris, R. H. T. P. (1921) A note on *Ortalia pallens* Muls. *South African Journal of Science* **17**, 170–171.

Harvey, P. H. & Pagel, M. D. (1991). *The Comparative Method in Evolutionary Biology*. Oxford: Oxford University Press.

Harvey, P. H., Read, A. F. & Promislow, D. E. L. (1989) Life history variation in placental mammals: unifying the data with theory. *Oxford Survey of Evolutionary Biology* **6**, 13–31.

Hassell, M. P. (1978) *The Dynamics of Arthropod Predator–Prey Systems*. Princeton, NJ: Princeton University Press.

Hassell, M. P. (1992) What is searching efficiency? *Annals of Applied Biology* **101**, 170–175.

Hassell, M. P., Lawton, J. H. & Beddington, J. R. (1976) The components of arthropod predation. I. The prey death rate. *Journal of Animal Ecology* **45**, 135–164.

Hassell, M. P. & Southwood, T. R. E. (1978) Foraging strategies of insects. *Annual Review of Ecology and Systematics* **9**, 75–98.

Hassell, M. P. & Varley, G. C. (1969) New inductive population model for insect parasites and its bearing on biological control. *Nature* **223**, 1133–1137.

Hatch, M. H. (1961) *The Beetles of the Pacific Northwest Part III. Pselaphidae and Diversicornia.* Seattle, WA: University of Washington Press.

Hattingh, V. & Samways, M. J. (1993) Evaluation of artificial diets and two species of natural prey as laboratory food for *Chilocorus* spp. *Entomologia Experimentalis et Applicata* **69**, 13–20.

Hattingh, V. & Samways, M. J. (1995) Visual and olfactory location of biotopes, prey patches, and individual prey by the ladybeetle *Chilocorus nigritus. Entomologia Experimentalis et Applicata* **75**, 87–98.

Heathcote, G. D. (1970) The abundance of grass aphids in Eastern England as shown by sticky trap catches. *Plant Pathology* **19**, 87–90.

Heathcote, G. D. (1978) Coccinellid beetles on sugar beet in Eastern England. *Plant Pathology* **27**, 103–109.

Hecht, O. (1936) Studies on the biology of *Chilocorus bipustulatus* (Coleoptera – Coccinellidae) an enemy of the red scale *Chrysomphalus auratii. Bulletin Society Fouad 1er Entomology* **20**, 299–326.

Heidari, M. & Copland, M. J. W. (1992) Host finding by *Cryptolaemus montrouzieri* (Col., Coccinellidae) a predator of mealybugs (Hom., Pseudococcidae). *Entomophaga* **37**, 621–625.

Heinz, K. M., Brazzle, J. R., Parrella, M. P. & Pickett, C. H. (1999) Field evaluations of augmentative releases of *Delphastus catalinae* Horn (Coleoptera : Coccinellidae) for suppression of *Bemisia argentifolii* Bellows & Perring (Homoptera : Aleyrodidae) infesting cotton. *Biological Control*

Heinz, K. M., Brazzle, J. R., Pickett, C. H., Natwick, E. T., Nelson, J. M. & Parrella, M. P. (1994) Predatory beetle may suppress silverleaf whitefly. *California Agriculture* **48**, 35–40.

Heinz, K. M. & Parrella, M. P. (1994) Poinsettia (*Euphorbia pulcherrima* Willd. ex Koltz.) cultivar-mediated differences in performance of five natural enemies of *Bemisia argentifolii* Bellows and Perring, n. sp. (Homoptera: Aleyrodidae). *Biological Control* **4**, 305–308.

Heinz, K. M. & Zalom, F. G. (1996) Performance of the predator *Delphastus pusillus* on *Bemisia* resistant and susceptible tomato lines. *Entomologia Experimentalis et Applicata* **81**, 345–352.

Hemptinne, J.-L., Dixon, A. F. G. & Adam, B. (2000a) Synchronization of sexual maturity in males and females of the two-spot ladybird *Adalia bipunctata* (L.) (Coleoptera: Coccinellidae). *Behavioral Ecology and Sociobiology*

Hemptinne, J.-L., Dixon, A. F. G. & Coffin, J. (1992) Attack strategy of ladybird beetles (Coccinellidae): factors shaping their numerical response. *Oecologia* **90**, 238–245.

Hemptinne, J.-L., Lognay, G. & Dixon, A. F. G. (1998) Mate recognition in the two-spot ladybird beetle, *Adalia bipunctata*: role of chemical and behavioural cues. *Journal of Insect Physiology* **44**, 1163–1171.

Hemptinne, J.-L., Dixon, A. F. G. & Gauthier, C. (2000b) Cost of intraguild predation in ladybirds (Coleoptera: Coccinellidae). *Ecological Entomology*

Hemptinne, J.-L., Doumbia, M. & Dixon, A. F. G. (2000c) Assessment of patch quality by ladybirds: role of aphids and plant phenology. *Journal of Insect Behavior*

Hemptinne, J.-L., Gaudin, M., Dixon, A. F. G. & Lognay, G. (2000d) Social feeeding in ladybird beetles: adaptive significance and mechanism. *Ecological Entomology*

Hemptinne, J.-L., Lognay, G., Doumbia, M. & Dixon, A. F. G. (2000e) Chemical nature of the oviposition deterring pheromone of the ladybird, *Adalia bipunctata*. *Proceedings of the Royal Society*

Hippa, H., Koponen, S. & Roine, R. (1982) Feeding preference of *Coccinella hieroglyphica* (Col., Coccinellidae) for eggs of three chrysomelid beetles. *Report of the Kevo Subarctic Research Station* **18**, 1–4.

Hironori, Y. & Katsuhiro, S. (1997) Cannibalism and interspecific predation in two predatory ladybirds in relation to prey abundance in the field. *Entomophaga* **42**, 153–163.

Hodek, I. (1959) Ecology of aphidophagous Coccinellidae. In *International Conference on Insect Pathology and Biological Control, Prague (1958)*, pp. 543–547.

Hodek, I. (1962) Essential and alternative food in insects. In *11th International Congress of Entomology, Vienna (1960)*, vol. 2, pp. 696–697.

Hodek, I. (1967) Bionomics and ecology of predaceous Coccinellidae. *Annual Review of Entomology* **12**, 79–104.

Hodek, I. (1970) Coccinellids and the modern pest management. *Bioscience* **20**, 543–552.

Hodek, I. (1973) *Biology of Coccinellidae*. Prague: Academia.

Hodek, I. (1993) Prey and habitat specificity in aphidophagous predators (a review). *Biocontrol Science and Technology* **3**, 91–100.

Hodek, I. & Cerkasov, J. (1960) Prevention and artificial induction of imaginal diapause in *Coccinella 7-punctata* L. *Nature* **187**, 345.

Hodek, I. & Cerkasov, J. (1963) Imaginal dormancy in *Semiadalia undecimnotata* Schneid. (Coccinellidae, Col.) II. Changes in water, fat and glycogen content. *Vestnik Ceskoslovenske Spolecnosti Zoologicke* **27**, 298–318.

Hodek, I., Chakrabarti, S. & Rejmanek, M. (1984a) The effect of prey density on food intake by adult *Cheilomenes sulphurea* [Col.: Coccinellidae]. *Entomophaga* **29**, 179–184.

Hodek, I. & Honěk, A. (1996) *Ecology of Coccinellidae*. Dordrecht: Kluwer Academic Publishers.

Hodek, I., Okuda, T. & Hodkova, M. (1984b) Reverse photoperiodic responses in two subspecies of *Coccinella septempunctata* L. *Zool Jb Syst* **111**, 439–448.

Hoelmer, K. A., Osborne, L. S. & Yokomi, R. K. (1993) Reproduction and feeding behaviour of *Delphastus pusillus* (Coleoptera : Coccinellidae) a predator of *Bemisia tabaci* (Homoptera : Aleyrodidae). *Journal of Economic Entomology* **86**, 322–329.

Holling, C. S. (1959) The components of predation as revealed by a study of small mammal predation of the european pine sawfly. *Canadian Entomologist* **91**, 293–320.

Holling, C. S. (1966) The functional response of invertebrate predators to prey density. *Memoirs of the Entomological Society of Canada* **48**, 1–86.

Holloway, G. J., De Jong, P. & Ottenheim, M. (1993) The genetics and cost of chemical defense in the two-spot ladybird (*Adalia bipunctata* L.). *Evolution* **47**, 1229–1239.

Holt, R. D. (1977) Predation, apparent competition and the structure of prey communities. *Theoretical Population Biology* **12**, 197–229.

Holt, R. D. & Hochberg, M. E. (1997) When is biological control evolutionarily stable (or is it?). *Ecology* **78**, 1673–1683.

Holt, R. D. & Lawton, J. H. (1994) The ecological consequences of shared natural enemies. *Annual Review of Ecology and Systematics* **25**, 495–520.

Honěk, A. (1978) Trophic regulation of postdiapause ovariole maturation in *Coccinella septempunctata* [Col.: Coccinellidae]. *Entomophaga* **23**, 213–216.

Honěk, A. (1985) Habitat preferences of aphidophagous coccinellids [Coleoptera]. *Entomophaga* **30**, 253–264.

Honěk, A. (1989) Overwintering and annual changes of abundance of *Coccinella septempunctata* in Czechoslovakia (Coleoptera: Coccinellidae). *Acta Entomologica Bohemoslovaca* **86**, 179–192.

Honěk, A. (1997) Factors determining winter survival in *Coccinella septempunctata* (Col.: Coccinellidae). *Entomophaga* **42**, 119–124.

Hopper, K. R. (1998) Assessing and improving the safety of introductions for biological control. In *Proceedings of the 6th Australasian Applied Entomological Research Conference*, vol. 1, ed. M. P. Zalucki, R. A. I. Drew & G. G. White, pp. 501–510.

Houck, M. A. (1986) Prey preference of *Sethorus punctorum* (Coleoptera: Coccinellidae). *Environmental Entomologist* **15**, 967–970.

Houck, M. A. & Strauss, R. E. (1985) The comparative study of functional responses: experimental design and statistical interpretation. *Canadian Entomologist* **117**, 617–629.

Houston, K. J. (1988) Larvae of *Coelophora inaequalis* (F.), *Phrynocaria gratiosa* (Mulsant) and *P. astrolabiana* (Weise) (Coleoptera : Coccinellidae) with notes on their relationships and prey records. *Journal of the Australian Entomological Society* **27**, 199–211.

Howarth, F. G. (1983) Classical biocontrol: panacea or Pandora's box. *Proceedings of the Hawaiian Entomological Society* **24**, 239–244.

Howarth, F. G. (1991) Environmental impacts of classical biological control. *Annual Review of Entomology* **36**, 485–509.

Howland, D. E. & Hewitt, G. M. (1995) Phylogeny of the Coleoptera based on mitochondrial cytochrome oxidase 1 sequence data. *Insect Molecular Biology* **4**, 203–215.

Hoy, M. A. (1992) Biological control of arthropods: genetic engineering and environmental risks. *Biological Control* **2**, 166–170.

Hukusima, S. & Kamei, M. (1970) Effect of various species of aphids as food on development, fecundity and longevity of *Harmonia axyridis* Pallas (Coleoptera: Coccinellidae). *Research Bulletin of the Faculty of Agriculture of Gifu University* **29**, 55–66.

Hukusima, S. & Komada, N. (1972) Longevity and fecundity of overwintered adults of *Propylea japonica*. *Research Bulletin of the Faculty of Agriculture of Gifu University* **33**, 83–87.

Hukusima, S. & Kouyama, S. (1974) Life histories and food habits of *Menochilus sexmaculatus* Fabricius (Coleoptera : Coccinellidae). *Research Bulletin of the Faculty of Agriculture of Gifu University* **36**, 19–29.

Hukusima, S. & Ohwaki, T. (1972) Further notes on feeding biology of *Harmonia axyridis* Pallas. *Research Bulletin of the Faculty of Agriculture of Gifu University* **33**, 75–82.

Hunter, K. W. (1978) Searching behaviour of *Hippodamia convergens* larvae (Coccinellidae: Coleoptera). *Psyche* **85**, 249–253.

Hurst, G. D. D., Sloggett, J. J. & Majerus, M. E. N. (1996) Estimation of the rate of inbreeding in a natural population of *Adalia bipunctata* (Coleoptera: Coccinellidae) using a phenotypic indicator. *European Journal of Entomology* **93**, 145–150.

Hutchinson, G. E. (1959) Homage to Santa Rosalia, or why are there so many kinds of animals? *American Naturalist* **93**, 145–159.

Iablokoff-Khnzorian, S. M. (1971). Synopsis des *Hyperaspis* Paléartiques (Coleoptera: Coccinellidae). *Annales de la Société entomologique de France (N.S.)* **7**, 163–200.

Iablokoff-Khnzorian, S.M. (1982). *Les Coccinelles – Coléoptères, Coccinellidae, tribu Coccinellini des régions Paléarctique et Orientale*. Paris: Boubée.

Iperti, G. & Buscarlet, L. A. (1986) Seasonal migration of the ladybird *Semiadalia undecimnotata*. In ed. I. Hodek, *Ecology of Aphidophaga*, vol. 2. pp. 199–204. Prague: Academia.

Iperti, G., Katsoyannos, P. & Laudeho, Y. (1977) Étude comparative de l'anatomie des coccinelles aphidiphages et coccidiphages et appartenance d'*Exochomus quardipustulatus* L, à l'un de ces groupes entomophages (Col. Coccinellidae). *Annales de la Société entomologique de France (N.S.)* **13**, 427–437.

Iwata, K. (1932a). On the biology of two large ladybirds in Japan. *Miscellaneous Contributions Entomological Laboratory of Kyoto Imperial University* **28**, 13–26.

Iwata, K. (1932b). On the biology of two large ladybirds in Japan. *Transactions of the Kansai Entomological Society* **3**, 13–26.

Jalali, S. K. & Singh, S. P. (1989a) Release and recovery of an exotic coccinellid predator, *Curinus coeruleus* (Muls.) on subabul(?) psyllid, *Heteropsylla cubana* (Crawf.) in India. *Journal of Insect Science* **2**, 158–159.

Jalali, S. K. & Singh, S. P. (1989b) Biotic potential of three coccinellid predators on various Diaspine hosts. *Journal of Biological Control* **3**, 20–23.

Janssen, A. & Sabelis, M. W. (1992) Phytoseiid life-histories, local predator–prey dynamics, and strategies for control of tetranychid mites. *Experimental and Applied Acarology* **14**, 233–250.

Jarosik, V., Honek, A. & Dixon, A. F. G. (2000) Rate isomorphy in insects and mites.

Jeffries, J. J. & Lawton, J. H. (1984) Enemy-free space and the structure of ecological communities. *Biological Journal of the Linnean Society* **23**, 269–286.

Jeffries, J. J. & Lawton, J. H. (1985) Predator–prey ratios in communities of freshwater invertebrates: the role of enemy-free space. *Freshwater Biology* **15**, 105–112.

Jervis, M. A. (1997) Parasitoids as limiting and selective factors: can biological control be evolutionarily stable? *Trends in Ecology and Evolution* **12**, 378–379.

Johnson, D. M., Akre, B. G. & Crowley, P. H. (1975) Modelling arthropod predation: wasteful killing by damselfly naiads. *Ecology* **56**, 1081–1093.

Jones, T. H. & Blum, M. S. (1983) Arthropod alkaloids: distribution, functions, and chemistry. In *Alkaloids. Vol. 1, Chemical and Biological Perspectives*, ed. S. W. Pelletier, pp. 33–84. New York: John Wiley.

de Jong, P. W., Holloway, G. J., Brakefield, P. M. & De Vos, H. (1991) Chemical defence in ladybird beetles (Coccinellidae) II. Amount of reflex fluid, the alkaloid adaline and individual variation in defence in 2-spot ladybirds (*Adalia bipunctata*). *Chemoecology* **2**, 15–19.

Kaddou, I. K. (1960) The feeding behavior of *Hippodamia quinquesignata* (Kirby) larvae. *University of California Publications in Entomology* **16**, 181–230.

Kairo, M. T. K. & Murphy, S. T. (1995) The life history of *Rodolia iceryae* Janson (Col., Coccinellidae) and the potential for use in innoculative releases against *Icerya pattersoni* Newstead (Hom., Margarodidae) on coffee. *Journal of Applied Entomology* **119**, 487–491.

Kalushkov, P. (1998) Ten species (Sternorrhyncha: Aphididae) as prey for *Adalia bipunctata* (Coleoptera: Coccinellidae). *European Journal of Entomology* **95**, 343–349.

Kalushkov, P. (1999) The effect of aphid prey quality on searching behaviour of *Adalia bipunctata* and its susceptibility to insecticides. *Entomologia Experimentalis et Applicata* **92**, 277–282.

Kamiya, H. (1961a) A revision of the tribe Scymnini from Japan and the Loochoos (Coleoptera: Coccinellidae), Part 1 Genera *Clitostethus*, *Stethorus* and *Scymnus* (except subgenus *Pullus*). *Journal of the Faculty of Agriculture of Kyushu University* **11**, 275–301.

Kamiya, H. (1961b) A revision of the tribe Scymnini from Japan and the Loochoos (Coleoptera: Coccinellidae), Part II Genus *Scymnus* (subgenus *Pullus*). *Journal of the Faculty of Agriculture of Kyushu University* **11**, 303–330.

Kamiya, H. (1965) A revision of the tribe Coccinellini of Japan and the Ryukyus (Coleoptera: Coccinellidae). *Memoirs of the Faculty of Liberal Arts of Fukui University, Series II Natural History* **15**, 27–71.

Kamiya, H. (1966). On the Coccinellidae attacking the scale insects and mites in Japan and Ryukyus. *Mushi* **39**, 66–93.

Kan, E. (1988a) Assessment of aphid colonies by hoverflies, I. Maple aphids and *Episyrphus balteatus* (de Geer)(Diptera: Syrphidae). *Journal of Ethology* **6**, 39–48.

Kan, E. (1988b) Assessment of aphid colonies by hoverflies, II. Pea aphids and three syrphid species: *Betasyrphus serarius* (Wiedemann), *Metasyrphus frequens* (Matsumara) and *Syrphus vitripennis* (Meigen) (Diptera: Syrphidae). *Journal of Ethology* **6**, 135–142.

Kan, E. & Sasakawa, M. (1986) Assessment of maple aphid colony by the hoverfly, *Episyrphus balteatus* (de Geer) (Diptera: Syrphidae). *Journal of Ethology* **4**, 121–127.

Kanika-Kiamfu, A., Kiyindou, A., Brun, J. & Iperti, G. (1992) Comparison des potentialités biologiques de trois coccinelles prédatrices de la cochenille farineuse du manioc *Phenacoccus manihoti* (Hom. Pseudococcidae). *Entomophaga* **37**, 277–282.

Kapur, A. P. (1942) Bionomics of some Coccinellidae, predaceous on aphids and coccids in North India. *Indian Journal of Entomology* **4**, 49–66.

Kareiva, P. (1986) Trivial movement and foraging by crop colonizers. In *Ecological Theory and Integrated Pest Practice*, ed. M. Kogan, pp. 59–82. New York: John Wiley.

Kareiva, P. (1996) Contribution of ecology to biocontrol. *Ecology* **77**, 1964–1965.

Kareiva, P. & Odell, G. (1987) Swarms of predators exhibit 'preytaxis' if individual predators use area-restricted search. *American Naturalist* **130**, 233–270.

Kareiva, P. & Sahakian, R. (1990) Tritrophic effects of a simple architectural mutation in pea plants. *Nature* **345**, 433–434.

Kariluoto, K. T. (1980) Survival and fecundity of *Adalia bipunctata* (Coleoptera, Coccinellidae) and some other predatory insect species on an arificial diet and a natural prey. *Annales Entomologici Fennici* **46**, 101–106.

Kaspari, M. (1996) Worker size and seed size selection by harvester ants in a neotropical forest. *Oecologia* **105**, 397–404.

Kawai, A. (1976) Analysis of the aggregation behaviour in the larvae of *Harmonia axyridis* pallas (Coleoptera : Coccinellidae) to prey colony. *Researches in Population Ecology* **18**, 123–134.

Kawai, A. (1978) Sibling cannibalism in the first instar larvae of *Harmonia axyridis* (Coleoptera, Coccinellidae). *Kontyu* **46**, 14–19.

Kawauchi, S. (1979) Effects of prey density on the rate of prey consumption, development and survival of *Propylea japonica* Thunberg (Coleoptera: Coccinellidae). *Kontyu* **47**, 204–212.

Kawauchi, S. (1981) The number of oviposition, hatchability and the term of oviposition of *Propylea japonica* Thunberg (Coleoptera, Coccinellidae) under different food conditions. *Kontyu* **49**, 183–191.

Kawauchi, S. (1983) The threshold temperature and thermal constant for development from egg to the adult form of *Coccinella septempunctata bruckii*, *Propylea japonica* and *Scymnus (Pullus) hoffmani* (Coleoptera, Coccinellidae). *Kurum University Journal* **32**, 45–51.

Kawauchi, S. (1985) Comparative studies on the fecundity of three aphidophagous coccinellids (Coleoptera: Coccinellidae). *Japanese Journal of Applied Entomology and Zoology* **29**, 203–209.

Kawauchi, S. (1995) Effects of photoperiod on the induction of diapause, the live weight of emerging adult and the duration of development of three species of aphidophagous coccinellids (Coleoptera, Coccinellidae). *Kontyu* **53**, 536–546.

Kaylani, S. (1967) Biology and life history of *Stethorus gilvifrons* Mulsant in Lebanon. *Magon Institut Recherche agronomique de Liban* **11**, 1–24.

Kehat, M. (1967) Studies on the biology and ecology of *Pharoscymnus numidicus* [Coccinellidae] an important predator of the date palm scale *Parlatoria blanchardi*. *Annales de la Société entomologique de France (N.S.)* **3**, 1053–1065.

Kehat, M. (1968) The feeding behaviour of *Pharoscymnus numidicus* (Coccinellidae), predator of the date palm scale *Parlatoria blanchardi*. *Entomologia Experimentalis et Applicata* **11**, 30–42.

Kesten, U. (1969). Zur Morphologie und Biologie von *Anatis ocellata* (L.) (Coleoptera, Coccinellidae). *Zeitschrift für angewandte Entomologie* **63**, 412–445.

Khalil, S. K., Shah, M. A. & Baloch, U. K. (1985) Optical orientation in predatory coccinellids. *Pakistan Journal of Agricultural Research* **6**, 40–44.

Kindlmann, P. & Dixon, A. F. G. (1992) Optimum body size: effects of food quality and temperature, when reproductive growth rate is restricted, with examples from aphids. *Journal of Evolutionary Biology* **5**, 677–690.

Kindlmann, P. & Dixon, A. F. G. (1993) Optimal foraging in ladybird beetles (Coleoptera: Coccinellidae) and its consequences for their use in biological control. *European Journal of Entomology* **90**, 443–450.

Kindlmann, P. & Dixon, A. F. G. (1999a) Generation time ratios – determinants of prey abundance in insect predator–prey interactions. *Biological Control* **16**, 133–138.

Kindlmann, P. & Dixon, A. F. G. (1999b) Stagies of aphidophagous predators: lessons for modelling insect predator–prey dynamics. *Journal of Applied Entomology* **123**, 397–399.

Kindlmann, P., Dixon, A. F. G. & Dostálková, I. (1999) Does body size optimization result in skewed body size distribution on a logarithmic scale? *American Naturalist* **153**, 445–447.

King, A. G. & Meinwald, J. (1996) Review of the defensive chemistry of coccinellids. *Chemical Review* **96**, 1105–1122.

Kirby, R. D. & Ehler, L. F.(1977) Survival of *Hippodamia convergens* in grain sorghum. *Environmental Entomology* **6**, 777–780.

Kirby, W. & Spence, W. (1815) *An Introduction to Entomology*. London: Longman, Brown, Green & Longman.

Kirk, W. D. J. (1991). The size relationship between insects and their hosts. *Ecological Entomology* **16**, 351–359.

Kislow, C. J. & Edwards, L. J. (1972) Repellent odour in aphids. *Nature* **235**, 108–109.

Klausnitzer, B. (1993) Zur Nahrungsökologie der mitteleuropäischen Coccinellidae (Col.). *Jahrbücher naturwissenschftlicher Verlage Wuppertal* **46**, 15–22.

Klausnitzer, B. & Klausnitzer, H. (1997) *Marienkäfer Coccinellidae*. Magdeburg: Westarp Wissenschaften.

Klingauf, F. (1967) Abwehr- und Meideraktionen von Blattläusen (Aphididae) bei Bedrohung durch Räubern und Parasiten. *Zeitschrift für angewandte Entomologie* **60**, 269–317.

Koide, T. (1962) Observations on the feeding habit of the larva of *Coccinella septempunctata bruckii* Mulsant. The feeding behaviour and number of prey fed under different temperatures. *Kontŷ* **30**, 236–241.

Kozar, F. & Walter, J. (1985). Check-list of the Palearctic Coccoïdae (Homoptera). *Folia Entomologica Hungarica* **46**, 63–110.

Kozlowski, J. & Weiner, J. (1997) Interspecific allometries are by-products of body size optimisation. *American Naturalist* **149**, 352–380.

Krebs, J. R. (1978) Optimal foraging: decision rules for predators. In *Behavioural Ecology, an Evolutionary Approach*, ed. J. R. Krebs & N. B. Davies, pp. 23–63. Oxford: Blackwell Scientific Publications, Oxford.

Kreiter, S. & Iperti, G. (1984) Étude des potentialités biologiques et écologiques d'un prédateur aphidophage *Olla v-nigrum* Muls. (Coleoptera, Coccinellidae) en vue de son introduction en France. In *109ᵉ Congrès national des Sociétés savantes, Dijon, Sciences*, vol. 2, pp. 275–282.

Kumashiro, B. R., Lai, P. Y., Funasaki, G. Y. & Teramoto, K. K. (1983) Efficacy of *Nephaspis ?haitiensis* in controlling *Aleurodiscus dispersus* in Hawaii. *Proceedings of the Hawaiian Entomological Society* **24**, 261–269.

Kuschel, G. (1990) *Beetles in a Suburban Environment: A New Zealand Case Study*. Department of Scientific and Industrial Research Plant Protection Report No. 3.

Lack, D. (1947) *Darwin's Finches*. Cambridge: Cambridge University Press.

Lack, D. (1986) *Ecological Adaptations for Breeding Birds*. London: Methuen.

Lambin, M., Ferran, A. & Maugan, K. (1996) La prise d'informations visuelles chez la coccinelle *Harmonia axyridis*. *Entomologia Experimentalis et Applicata* **79**, 121–130.

Lane, S. D., Mills, N. J. & Getz, W. M. (1999) The effects of parasitoid fecundity and taxon on the biological control of insect pests: the relationship between theory and data. *Ecological Entomology* **24**, 181–190.

Lawton, J. H. (1986) The effects of parasitoids on phytophagous insect communities. In *Insect Parasitoids*, ed. J. K. Waage & D. Greathead, pp. 265–287. London: Academic Press.

Lawton, J. H., Hassell, M. P. & Beddington, J. R. (1975) Prey death rates and rate of increase of arthropod predator populations. *Nature* **255**, 60–62.

Leeper, J. R. & Beardsley, J. W. (1976) The biological control of *Psylla uncatoides* (Ferris & Klyver) (Homoptera: Psyllidae) on Hawaii. *Proceedings of the Hawaiian Entomological Society* **22**, 307–321.

Liepert, C. & Dettner, K. (1996) Role of cuticular hydrocarbons of aphid parasitoids in their relationship to aphid-attending ants. *Journal of Chemical Ecology* **22**, 695–707.

Liu, C.-Y. (1933) Notes on the biology of two giant coccinellids in Kwangsi (*Caria dilatata* Fabr. and *Synonycha grandis* Thunbg.) with special reference to the morphology of *Caria dilatata*. *Year Book of the Bureau of Entomology, Hangchow* **1**, 205–250.

Liu, T.-X., Stansly, P. A., Hoelmer, K. A. & Osborne, L. S. (1998) Life history of *Nephaspis aculatus* (Coleoptera: Coccinellidae), a predator of *Bemisia argentifolii* (Homoptera: Aleyrodidae). *Annals of the Entomological Society of America* **90**, 776–782.

Longo, S. & Benfatto, D. (1987). Coleotteri entomofagi presenti sugli agrumi in Italia. *Informatore Fitopatologico* **7–8**, 21–30.

Longworth, J. F. (1987) Biological control in New Zealand: policy and procedures. *New Zealand Entomologist* **10**, 1–7.

Losey, J. E. & Denno, R. F. (1998) Positive predator–predator interactions: enhanced predation rates and synergistic suppression of aphid populations. *Ecology* **79**, 2143–2152.

Lounsbury, C. P. (1940) The pioneer period of economic entomology in South Africa. *Journal of the Entomological Society of South Africa* **3**, 9–29.

Lucas, E., Coderre, D. & Brodeur, J. (1997a) Instar-specific defense of *Coleomegilla maculata lengi* (Col.: Coccinellidae): influence on attack success of the intra-guild predator *Chrysoperla rufilabris* (Neur.: Chrysopidae). *Entomophaga* **42**, 3–12.

Lucas, E., Coderre, D. & Brodeur, J. (1998) Intraguild predation among aphid predators: characterization and influence of extraguild prey density. *Ecology* **79**, 1084–1092.

Lucas, E., Coderre, D. & Vincent, C. (1997b) Voracity and feeding preference of two aphidophagous coccinellids fed on *Aphis citricola* and *Tetranychus urticae*. *Entomologia Experimentalis et Applicata* **85**, 151–159.

Luck, R. F. (1990) Evaluation of natural enemies for biological control: a behavioral approach. *Trends in Ecology and Evolution* **5**, 196–199.

Mader, L. (1926–37). Evidenz der paläarktischen Coccinelliden und ihrer Aberrationen, In Wort und Bild, I. Epilachnini, Coccinellini, Halyziini, Synonychini. (1926–34) *Zeitschrift des Vereins Naturbeobachter und Sammler;* (1935) *Entomologischer Anzeiger* **15**, 329–383; (1937) *Entomologischer Nachrichtenblatt* 384–412.

Mader, L. (1941) Coccinellidae I. Teil. *Exploration du parc national Albert*, Mission G.F. de Witte (1933–1935) **34**, 1–208.

Mader, L. (1950) Coccinellidae II. Teil. *Exploration du parc national Albert*, Mission G.F. de Witte (1933–1935) **43**, 1–134.

Mader, L. (1954) Coccinellidae III. Teil. *Exploration du parc national Albert*, Mission G.F. de Witte (1933–1935) **47**, 1–206.

Mader, L. (1955) Evidenz der paläarktischen Coccinelliden un ihrer Aberrationen in Wort und Bild, II. Teil, Coccinellidae, Tetrabrachinae. *Entomologische Arbeiten Museum G. Frey Tutzing bei München* **6**, 764–1035.

Maelzer, D. A. (1978) The growth and voracity of larvae of *Leis conformis* (Boisd.) (Coleoptera : Coccinellidae) fed on the rose aphid *Macrosiphum rosae* (L.) (Homoptera : Aphididae) in the laboratory. *Australian Journal of Zoology* **26**, 293–304.

Magro, A. (1997) Os Coccinelídeos dos citrinos: estudo comparativo do seu interesse em lutta biológica. PhD thesis, Universidade de Évora, Portugal.

Majerus, M. E. N. (1989) *Coccinella magnifica* (Redtenbacher): a myrmecophilous ladybird. *British Journal of Entomology and Natural History* **2**, 43–52.

Majerus, M. E. N. (1994) *Ladybirds*. London: HarperCollins.

Majerus, M. E. N. (1998) The private lives of British ladybirds. *British Wildlife* **998**, 233–242.

Majerus, M. E. N. & Hurst, G. D. D. (1997) Ladybirds as a model system for the study of male-killing symbionts. *Entomophaga* **42**, 13–20.

Majerus, M. E. N. & Majerus, T. M. O. (1996) Ladybird population explosions. *British Journal of Entomology and Natural History* **9**, 65–76.

Majerus, M. E. N. & Majerus, T. M. O. (1998) Mimicry in ladybirds. *Bulletin of the Amateur Entomologists' Society* **57**, 126–140.

Malcolm, S. B. (1992) Prey defence and predator foraging. In *Natural Enemies: The Population Biology of Predators, Parasites and Diseases*, M. J. Crawley, pp. 458–475. Oxford: Blackwell Scientific Publications.

Mangel, M. (1987) Oviposition site selection and clutch size in insects. *Journal of Mathematical Biology* **25**, 1–22.

Mani, M. & Thontadarya, T. S. (1987) Biological studies on the grapevine mealybug predator, *Scymnus coccivora* Ayar (*Coccinellidae: Coleoptera*). *Journal of Biological Control* **1**, 89–92.

Marcovitch, S. (1935) Experimental evidence on the value of strip farming as a method for the natural control of injurious insects with special reference to plant lice. *Journal of Economic Entomology* **28**, 62–70.

Marin, R. (1983) Biología y capacidad de predación de *Lindorus lophanthae* (Blais) (Col.: Coccinellidae) predator de *Pinnaspis aspidistrae* (Sing) (Hom.: Diaspididae). *Revista Peruana de Entomología* **26**, 63–66.

Marks, R. J. (1977) Laboratory studies of plant searching by *Coccinella septempunctata* L. larvae. *Bulletin of Entomological Research* **67**, 235–241.

Marples, N. M. (1990) Influence of predation on ladybird colour pattern. PhD thesis, University of Wales College of Cardiff, U. K.

Marples N. M., Brakefield, P. M. & Cowie, R. J. (1989) Differences between the 7-spot and 2-spot ladybird beetles (Coccinellidae) in their toxic effects on a bird predator. *Ecological Entomology* **14**, 79–84.

Marples, N. M., De Jong, P. W., Ottenheim, M. M., Verhoog, M. D. & Brakefield, P. M. (1993) The inheritance of a wingless character in the 2 spot ladybird (*Adalia bipunctata*). *Entomologia Experimentalis et Applicata* **69**, 69–73.

Matsuka, M., Watanabe, M. & Niijima, K. (1982) Longevity and oviposition of vedalia beetles on artificial diets. *Environmental Entomology* **11**, 816–819.

Matsura, T. (1976) Ecological studies of a coccinellid, *Aiolocaria hexaspilota* Hope I. Interaction between field populations of *A. hexaspilota* and its prey, the walnut leaf beetle (*Gastrolina depressa* Baly). *Japanese Journal of Ecology* **26**, 147–156.

Maudsley, M. J. (1993) Regional differences in the abundance of cereal aphids. PhD thesis, University of East Anglia, U.K.

Maurer, B. A., Brown, J. H. & Rusler, R. D. (1992). The micro and macro in body size evolution. *Evolution* **46**, 939–953.

May, R. M. & Hassell, M. P. (1981)The dynamics of multiparasitoid–host interactions. *American Naturalist* **117**, 234–261.

McKenzie, H. L. (1932) The biology and feeding habits of *Hyperaspis lateralis* Mulsant (Coleoptera – Coccinellidae). *University of California Publications in Entomology* **6**, 9–21.

McLean, I. F. G. (1980) Ecology of the natural enemies of cereal aphids. PhD thesis, University of East Anglia, U.K.

Medawar, P. B. (1965) Two conceptions of science. *Encounter* **143**.

van den Meiracker, R. A. F., Hammond, W. N. O. & van Alphen, J. J. M. (1990) The role of kairomones in prey finding by *Diomus* sp. and *Exochomus* sp., two coccinellid predators of the cassava mealybug, *Phenacoccus manihoti*. *Entomologia Experimentalis et Applicata* **56**, 209–217.

Mendel, Z., Podoler, H. & Rosen, D. (1984) Population dynamics of the Mediterranean black scale, *Saissetia oleae* (Olivier), on citrus in Israel. 4: The natural enemies. *Journal of the Entomological Society of South Africa* **47**, 1–21.

Merlin, J., Lemaitre, O. & Grégoire, J.-C. (1996a) Oviposition in *Cryptolaemus montrouzieri* stimulated by wax filaments of its prey. *Entomologia Experimentalis et Applicata* **79**, 141–146.

Merlin, J., Lemaitre, O. & Grégoire, J.-C. (1996b) Chemical cues produced by conspecific larvae deter oviposition by the coccidophagous ladybird beetle, *Cryptolaemus montrouzieri*. *Entomologia Experimentalis et Applicata* **79**, 147–151.

Messina, F. J. & Hanks, J. B. (1998) Host plant alters shape of the functional response of an aphid predator (Coleoptera : Coccinellidae). *Environmental Entomology* **27**, 1196–1202.

Messina, F. J., Jones, T. A. & Nielson, D. C. (1997) Host-plant effects on the efficacy of two predators attacking Russian wheat aphids (Homoptera: Aphididae). *Environmental Entomologist* **26**, 1398–1404.

Messing, R. H. & Duan, J. J. (1998) Effects of introduced fruit fly parasitoids on non-target tephritids in Hawaii. *Proceedings of the 6th Australasian Applied Entomological Research Conference*, vol. 1, ed. M. P. Zalucki, R. A. I. Drew & G. G. White, pp. 547–555.

Meyerdirk, D. E. (1983) Biology of *Diomus flavifrons* (Blackburn) (Coleoptera: Coccinellidae), a citrus mealybug predator. *Environmental Entomology* **12**, 1275–1277.

Michels, G. J. & Bateman, A. C. (1986) Larval biology of two imported predators of the greenbug , *Hippodamia variegata* Goetz and *Adalia flavomaculata* DeGeer, under constant temperatures. *The Southwestern Entomologist* **11**, 23–30.

Michels, G. J. & Behle, R. W. (1991) Effects of two prey species on the development of *Hippodamia sinuata* (Coleoptera: Coccinellidae) larvae at constant temperatures. *Journal of Economic Entomology* **84**, 1480–1484.

Miller, J. C. & LaMana, M. L.(1995) Assessment of temperature dependent development in the general population and among isofemale lines of *Coccinella trifasciata* (Col.: Coccinellidae). *Entomophaga* **40**, 183–192.

Mills, N. J. (1979) *Adalia bipunctata* (L) as a generalist predator of aphids. PhD thesis, University of East Anglia, U.K.

Mills, N. J. (1981a) Some aspects of the rate of increase of a coccinellid. *Ecological Entomology* **6**, 293–299.

Mills, N. J. (1981b) Essential and alternative foods for some British Coccinellidae (Coleoptera). *Entomologists' Gazette* **32**, 197–202.

Mills, N. J. (1982a) Voracity, cannibalism and coccinellid predation. *Annals of Applied Biology* **101**, 144–148.

Mills, N. J. (1982b) Satiation and the functional response: a test of a new model. *Ecological Entomology* **7**, 305–315.

Minelli, A. & Pasqual, C. (1977) The mouthparts of ladybirds: structure and function. *Bolletino Zoologia Napoli* **44**, 183–187.

Mineo, G. (1966) Sul *Cryptolaemus montrouzierri* Muls. (Osservazioni morfo-biologiche). *Bolletino Istituto di Entomologia Agraria dell' Università di Palermo* **6**, 99–143.

Moore, B. P. & Brown, W. V. (1981) Identification of warning odour components, bitter principles and antifeedants in an aposematic beetle: *Metriorrhynchus rhipidus* (Coleoptera: Lycidae). *Insect Biochemistry* **11**, 493–499.

Moore, B. P., Brown, W. V. & Rothschild, M. (1990) Methylalkylpyrazines in aposematic insects, their hostplants and mimics. *Chemoecology* **1**, 43–51.

Morales, J. & Burandt, C. L. (1985) Interactions between *Cycloneda sanguinea* and the brown citrus aphid: adult feeding and larval mortality. *Environmental Entomology* **14**, 520–522.

Morjan, W. E., Obrycki, J. J. & Krafsur, E. S. (1999) Inbreeding effects on *Propylea quatuordecimpunctata* (Coleoptera: Coccinellidae). *Annals of the Entomological Society of America* **92**, 260–268.

Morse, D. R., Lawton, J. H., Dodson, M. M. & Williamson, M. H. (1985) Fractal dimension of vegetation and the distribution of arthropod body lengths. *Nature* **314**, 731–733.

Moursi, A. A. & Kamal, M. (1946) Notes on the biology and feeding habits of the introduced beneficial insect *Leis conformis* Boisd. *Bulletin Society Fouad 1er Entomologie* **30**, 63–74.

Müller, C. B. & Godfray, H. C. J. (1997) Apparent competition between two aphid species. *Journal of Animal Ecology* **66**, 57–64.

Munyaneza, J. & Obrycki, J. J. (1998) Development of three populations of *Coleomegilla maculata* (Coleoptera: Coccinellidae) feeding on eggs of Colorado potato beetle (Coleoptera: Chrysomelidae). *Environmental Entomology* **27**, 117–122.

Murakami, Y. & Tsubaki, Y. (1984) Searching efficiency of the lady beetle *Coccinella septempunctata* larvae in uniform and patch environments. *Journal of Ethology* **2**, 1–6.

Murdoch, W. W. (1969) Switching in general predators: experiments on predator specificity and stability of prey populations. *Ecological Monographs* **39**, 335–354.

Murdoch, W. W. (1990) The relevance of pest–enemy models to biological control. In *Critical Issues in Biological Control*, ed. M. Mackauer, L. E. Ehler & J. Roland, pp. 1–24. Andover: Intecept Ltd.

Murdoch, W. W., Briggs, C. J. & Collier, T. R. (1998) Biological control of insects: implications for theory in population ecology. In *Insect Populations* ed. J. P. Dempster & I. F. G. McLean, pp. 167–186. Dordrecht: Kluwer Academic Publishers.

Murdoch, W. W. & Marks, R. J. (1973) Predation by coccinellid beetles: experiments on switching. *Ecology* **54**, 160–167.

Nadel, D. J. & Biron, S. (1964) Laboratory studies and controlled mass rearing of *Chilocorus bipustulatus* Linn., a citrus scale predator in Israel. *Rivista di Parasitologia* **25**, 195–206.

Nakamuta, K. (1984) Visual orientation of a ladybeetle, *Coccinella septempunctata* L.

(Coleoptera : Coccinellidae), towards its prey. *Applied Entomology and Zoology* **19**, 82–86.

Nakamuta, K. (1985a) Mechanism of the switchover from extensive to area-concentrated search behaviour of the ladybird beetle, *Coccinella septempunctata*. *Journal of Insect Physiology* **31**, 849–856.

Nakamuta, K. (1985b) Area-concentrated search in adult *Coccinella septempunctata* L. (Coleoptera : Coccinellidae): releasing stimuli and decision of giving-up time. *Japanese Journal of Applied Entomology and Zoology* **29**, 55–60.

Nakamuta, K. (1987) Diel rhythmicity of prey-search activity and its predominance over starvation in the lady beetle, *Coccinella septempunctata brucki*. *Physiological Entomology* **12**, 91–98.

Nakamuta, K. (1991) Aphid alarm pheromone component, (E)-β-farnesene, and local search by a predatory lady beetle, *Coccinella septempunctata brucki* Mulsant (Coleoptera: Coccinellidae). *Applied Entomology and Zoology* **26**, 1–7.

Nakamuta, K. & Saito, T. (1985) Recognition of aphid prey by the lady beetle, *Coccinella septempunctata brucki* Mulsant (Coleoptera : Coccinellidae). *Applied Entomology and Zoology* **20**, 479–483.

Napompeth, B. & Maneeratana, T. (1990) Biological and partial ecological life table of *Heteropsylla cubana* Crawford and its predator, *Curinus coeruleus* Mulsant in Thailand. In *Leucaena Psyllid: Problems and Management*, ed. B. Napompeth & K. G. MacDicken, pp. 130–138. Bogor, Indonesia.

Naranjo, S. E. , Gibson, R. L. & Walgenbach, D. D. (1990) Development, survival, and reproduction of *Scymnus frontalis* (Coleoptera : Coccinellidae), an imported predator of Russian wheat aphid, at four fluctuating temperatures. *Annals of the Entomological Society of America* **83**, 527–531.

Nechols, J. R. & Obrycki, J. J. (1989). Comparative behavioural and ecological studies in relation to biological control: an overview. *Journal of the Kansas Entomological Society* **62**, 146–147.

Nicholson, A. J. (1933) The balance of animal populations. *Journal of Animal Ecology* **2**, 132–178.

Nicholson, A. J. & Bailey, V. A. (1935) The balance of animal populations. *Proceedings of the Zoological Society of London* **1935**(3), 551–598.

Nicol, H. (1943) *The Biological Control of Insects*. Harmondsworth: Penguin Books.

Niijima, K., Matsuka, M. & Okada, I. (1986) Artificial diets for an aphidophagous coccinellid, *Harmonia axyridis*, and its nutrition (minireview). In *Ecology of Aphidophaga* ed I. Hodek, pp. 37–50. Prague: Academia.

Nsiama She, H. D., Odebiyi, J. A. & Herren, H. R. (1984) The biology of *Hyperaspis jucunda* (Col., Coccinellidae) an exotic predator of the cassava mealybug *Phenacoccus manihoti* (Hom., Pseudococcidae) in Southern Nigeria. *Entomophaga*, **29**, 87–93.

Nunnenmacher, L. (1998) Blattläuse auf Kopfsalat und deren Kontrolle durch gezielte Beeinflussung der Lebensgrundlagen ihrer Prädatoren. PhD thesis, Bayreuther Forum Ökologie Band 61.

Obata, S. (1986) Mechanisms of prey finding in the aphidophagous ladybird beetle, *Harmonia axyridis* [Coleoptera : Coccinellidae]. *Entomophaga* **31**, 303–311.

Obata, S. (1997) The influence of aphids on the behaviour of adults of the lady beetle, *Harmonia axyridis* (Col.: Coccinellidae). *Entomophaga* **42**, 103–106.

Obata, S. & Johki, Y. (1990) Distribution and behaviour of adult ladybird, *Harmonia axyridis* Pallas (Coleoptera, Coccinellidae), around aphid colonies. *Japanese Journal of Entomology* **58**, 839–845.

Obrycki, J. J. & Kring, T. T. (1998) Predaceous Coccinellidae in biological control. *Annual Review of Entomology* **43**, 295–321.

Obrycki, J. J., Orr, D. B., Orr, C. J., Wallendorf, M. & Flanders, R. V. (1993) Comparative development and reproductive biology of three populations of *Propylea quatu-ordecimpunctata* (Coleoptera: Coccinellidae). *Biological Control* **3**, 27–33.

Ofuya, T. I. (1995) Studies on the capability of *Cheilomenes lunata* (Fabricius) (Coleoptera: Coccinellidae) to prey on the cowpea aphid, *Aphis craccivora* (Koch) (Homoptera: Aphididae) in Nigeria. *Agricultural Ecosystems and Environment* **52**, 35–38.

Okamoto, H. (1978) Laboratory studies on food ecology of aphidophagous lady beetles (Coleoptera: Coccinellidae). *Memoirs of the Faculty of Agriculture of Kagaw University* **32**, 1–94.

Okrouhlá, M., Chakrababarti, S. & Hodek, I. (1983) Developmental rate and feeding capacity in *Cheilomenes sulphurea* (Coleoptera: Coccinellidae). *Vestnik Ceskoslovenske Spolecnosti Zoologicke* **47**, 105–117.

Osawa, N. (1989) Sibling and non-sibling cannibalism by larvae of a lady beetle *Harmonia axyridis* Pallas (Coleoptera: Coccinellidae) in the field. *Researches in Population Ecology* **31**, 153–160.

Osawa, N. (1992*a*) Sibling cannibalism in the lady beetle *Harmonia axyridis*: fitness consequence for mother and offspring. *Researches in Population Ecology* **34**, 45–55.

Osawa, N. (1992*b*) Effect of pupation site on pupal cannibalism and parsitism in the ladybird beetle *Harmonia axyridis* Pallas (Coleoptera, Coccinellidae). *Japanese Journal of Entomology* **60**, 131–135.

Osawa, N. (1992*c*) A life table of the ladybird beetle *Harmonia axyridis* Pallas (Coleoptera, Coccinellidae) in relation to aphid abundance. *Japanese Journal of Entomology* **60**, 575–579.

Osawa, N. (1993) Population field studies of the aphidophagous ladybird beetle *Harmonia axyridis* (Coleoptera; Coccinellidae): life tables and key factor analysis. *Researches in Population Ecology* **35**, 335–348.

Osawa, N. & Nishida, T. (1992) Seasonal variation in elytral colour polymorphism in *Harmonia axyridis* (the ladybird beetle): the role of non-random mating. *Heredity* **69**, 297–307.

Pagel, M. D., Harvey, P. H. & Godfray, H. C. J. (1991) Species-abundance, biomass, and resource-use distributions. *American Naturalist* **138**, 836–850.

Palmer, M. A. (1914) Some notes on life history of ladybeetles. *Annals of the Entomological Society of America* **7**, 213–238.

Pantyukhov, G. A. (1965) Influence of temperature and relative humidity on the development of *Chilocorus renipustulatus* Scriba, Coleoptera, Coccinellidae. *Trudy Zoologicheskii Instytute, Leningrad* **36**, 70–85.

Pantyukhov, G. A. (1968) A study of ecology and physiology of the predatory beetle *Chilocorus rubidus* Hope (Coleoptera, Coccinellidae). *Zoologicheskii Zhurnal* **47**, 376–386.

Parker, G. A. & Courtney, S. P. (1984) Models of clutch size in insect oviposition. *Theoretical Population Biology* **26**, 27–48.

Parry, W. H. (1980) Overwintering of *Aphidecta obliterata* (L.) (Coleoptera: Coccinellidae) in north east Scotland. *Acta Oecologica* **1**, 307–316.

Partridge, L. & Farquhar, M. (1981) Sexual activity reduces life span of male fruit flies. *Nature* **294**, 580–582.

Partridge, L. & Harvey, P. H. (1985) Cost of reproduction. *Nature* **316**, 20.

Pasteels, J.M. (1978) Apterous and brachypterous coccinellids at the end of the food chain, *Cionura erecta* (Asclepiadaceae) – *Aphis nerii*. *Entomologia Experimentalis et Applicata* **24**, 379–384.

Pasteels, J. M., Deroe, C., Tursch, B., Braekman, J. C., Daloze, D. & Hootele, C. (1973) Distribution et activités des alcaloides défensifs des Coccinellidae. *Journal of Insect Physiology* **19**, 1771–1784.

Patnaik, N. C. & Sahu, N. C. (1980) Influence of aphid host and certain artificial diets on ovipositional response and longevity of lady beetle, *Menochilus sexmaculatus* (Fabr.) (Coleoptera: Coccinellidae). *Science and Culture* **46**, 103–105.

Peters, R. H. (1983). *The Ecological Implications of Body Size*. Cambridge: Cambridge University Press.

Phoofolo, M. W. & Obrycki, J. J. (1997) Comparative prey suitability of *Ostrinia nubilialis* eggs and *Acyrthosiphon pisum* for *Coleomegilla maculata*. *Biological Control* **9**, 167–172.

Phoofolo, M. W. & Obrycki, J. J (1998) Potential for intraguild predation and competition among predatory Coccinellidae and Chrysopidae. *Entomologia Experimentalis et Applicata* **89**, 47–55.

Pienkowski, R. L. (1965) The incidence and effect of egg cannibalism in first-instar *Coleomegilla maculata lengi* (Coleoptera: Coccinellidae). *Annals of the Entomological Society of America* **58**, 150–153.

Pimentel, D., Genister, C., Fast, S. & Gallahan, D. (1984) Environmental risks of biological pest controls. *Oikos* **42**, 283–290.

Pimm, S. L. (1991) *The Balance of Nature: Ecological Issues in the Conservation of Species and Communities*. Chicago, IL: University of Chicago Press.

Podoler, H. & Henen, J. (1986) Foraging behaviour of two species of the genus *Chilocorus* (Coccinellidae: Coleoptera): a comparative study. *Phytoparasitica* **14**, 11–23.

Polis, G. A. (1981) The evolution and dynamics of intraspecific predation. *Annual Review of Ecology and Systematics* **12**, 225–251.

Polis, G. A. & Holt, R. D. (1992) Intraguild predation: the dynamics of complex trophic interactions. *Trends in Ecology and Evolution* **7**, 151–154.

Polis, G. A., Myers, C. A. & Holt, R. D. (1989) The ecology and evolution of intraguild predation: potential competitors that eat each other. *Annual Review of Ecology and Systematics* **20**, 297–330.

Ponsonby, D. J. & Copland, M. J. W. (1995) Olfactory responses by the scale insect predator *Chilocorus nigritus* (F.) (Coleoptera: Coccinellidae). *Biocontrol Science and Technology* **5**, 83–93.

Pope, R. D. (1953) *Coleoptera Coccinellidae & Sphindidae*. Royal Entomological Society of London Handbooks for the Identification of British Insects, vol. V, part 7.

Pope, R. D. (1973) The species of *Scymnus* (S.STR.), *Scymnus* (*Pullus*) and *Nephus* (Col., Coccinellidae) occurring in the British Isles. *Entomologists' Monthly Magazine* **109**, 3–39.

Pope, R. D. (1988) A revision of the Australian Coccinellidae (Coleoptera) Part 1. Subfamily Coccinellinae. *Invertebrate Taxonomy* **2**, 633–735.

Pope, R. D. & Lawrence, J. F. (1990) A review of *Scymnodes* Blackburn, with the description of a new Australian species and its larva (Coleoptera: Coccinellidae). *Systematic Entomology* **15**, 241–252.

Portevin, G. (1931). *Histoire naturelle des Coléoptères de France, Tome II. Polyphaga, Lamellicornia, Palpicornia, Diversicornia*. Paris: P. Chevalier.

Pradhan, S. (1936) The alimentary canal of *Epilachna indica* (Coccinellidae: Coleoptera) with a discussion on the activity of the mid-gut epithelium. *Journal of the Royal Asiatic Society of Bengal, Science* **2**, 127–156.

Pradhan, S. (1938) Neuro-muscular study of the mouth-parts of *Coccinella septempunctata*, with a comparison of the mouthparts in carnivorous and herbivorous coccinellids. *Records of the Indian Museum, Calcutta* **40**, 341–358.

Pradhan, S. (1939) The alimentary canal and pro-epithelial regeneration in *Coccinella septempunctata* with a comparison of carnivorous and herbivorous coccinellids. *Quarterly Journal of the Microscopical Science* **81**, 451–478.

Prakasan, C. B. & Bhat, K. P. (1985) Interference of the predator, *Cryptolaemus mon-*

trouzieri with the performance of a newly introduced parasite, *Leptomastix dactylopii. Journal of Coffee Research* **15**, 29–32.

Prasad, Y. K. (1990) Discovery of isolated patches of *Icerya purchasi* by *Rodolia cardinalis*: a field study. *Entomophaga* **35**, 421–429.

Promislow, D. E. L. & Harvey, P. H. (1990) Living fast and dying young: a comparative analysis of life-history variation among mammals. *Journal of Zoology, London* **220**, 417–437.

Pruszynski, S. & Lipa, J. J. (1970) Observations on life cycle and food specialization of *Adalia bipunctata* (L.) (Coleoptera, Coccinellidae). *Prace Naukowe Instytutu Ochrony Roslin* **12**, 99–116.

Pu, C. L. (1976) Biological control of insect pests in China. *Acta Entomologica Sinica* **19**, 247–252.

Putman, W.L. (1955) Bionomics of *Stethorus punctillum* Weise (Coleoptera: Coccinellidae) in Ontario. *Canadian Entomologist* **87**, 9–33.

Quezada, J. R. & DeBach, P. (1973(4)) Bioecological and population studies of the cottony-cushion scale, *Icerya purchasi* Mask., and its natural enemies, *Rodolia cardinalis* Mul. and *Cryptochaetum iceryae* Will., in Southern California. *Hilgardia* **41**, 631–688.

Quilici, S. (1981) Étude biologique de *Propylea quatuordecimpunctata* L. (Coleoptera, Coccinellidae). Efficacité prédatrice comparée de trois types de coccinelles aphidophages en lutte biologique contre les pucerons sous serre. PhD thesis, L'Université Pierre et Marie Curie, Paris VI, France.

Raghunath, T. A. V. S. & Rao, E. H. K. (1980) Studies on the biology and feeding potential of *Sticholotis madagassa* (Coccinellidae: Coleoptera) an exotic predator on the scale insect of sugarcane *Melanaspis glomerata* (Green). *Maharashtra Sugar*, June, 17–20.

Rajamohan, N. & Jayaraj, S. (1973) Growth and development of the coccinellid *Menochilus sexmaculatus* Fabricius on four species of aphid. *Madras Agricultural Journal* **61**, 118–122.

Raimundo, A. A. C. (1992) Novas espécies de Scymnini para a fauna de Coccinelideos de Portugal. *Boletim da Sociedade Portuguesa de Entomologia* suppl. no. **3**, 373–384.

Raimundo, A. A. C. & Alves, M. L. G. (1986) *Revisão dos Coccinelideos de Portugal.* Universidade de Évora.

Raupp, M. J., Hardin, M. R., Braxton, S. M. & Bull, B. B. (1994) Augmentative releases for aphid control on landscape plots. *Journal of Arboriculture* **20**, 241–249.

Reed, D. A. & Beckage, N. E. (1997) Inhibition of testicular growth and development in *Mandura sexta* larvae parasitized by the braconid wasp *Cotesia congregata. Journal of Insect Physiology* **43**, 29–38.

Reiss, M. J. (1989) *The Allometry of Growth and Reproduction.* Cambridge: Cambridge University Press, Cambridge.

Rhamhalinghan, M. (1985) Intraspecific variations in ovariole numbers/ovary in *Coccinella septempunctata* L. (Coleoptera : Coccinellidae). *The Indian Zoologist* **9**, 91–97.

Ricci, C. (1979) L'apparato boccale pungente succhiante della larva di *Platynaspis luteorubra* Goeze (Col. Coccinellidae). *Bollettino del Laboratorio di Entomologia Agraria 'Filippo Silvestri' di Portici* **36**, 179–198.

Ricci, C. (1982) Sulla costituzione e funzione delle mandible delle larve di *Tytthaspis sedecimpunctata* (L.) e *Tytthaspis trineleata* (Weise). *Frustula Entomologica* **3**, 205–212.

Ricci, C. & Cappelletti, G. (1988) Relationship between some morphological structures and locomotion of *Clitostethus arcuatus* Rossi (Coleoptera Coccinellidae), a whitefly predator. *Frustula Entomologica* **11**, 195–202.

Richards, D. R. & Evans, E. W. (1998) Reproductive responses of aphidophagous lady beetles (Coleoptera : Cocinellidae) to nonaphid diets: an example from alfalfa. *Annals of the Entomological Society of America* **91**, 632–640.

Rivnay, E. & Perzelan, J. (1943) Insects associated with *Pseudococcus* spp. (Homoptera) in Palestine, with notes on their biology and economic status. *Journal of the Entomological Society of South Africa* **6**, 9–28.

Rodriquez-Saonia, C. & Miller, J.C. (1995) Life history traits in *Hippodamia convergens* (Coleoptera: Coccinellidae) after selection for fast development. *Biological Control* **5**, 389–396.

Roff, D. A. (1992) *The Evolution of Life Histories*. New York: Chapman & Hall.

Roitberg, B. D. (1989) The cost of reproduction in rosehip flies, *Rhagoletis basiola*, eggs are time. *Evolutionary Ecology* **3**, 183–188.

Roland, J. (1998) The population dynamics of *Operophtera brumata* (Lepidoptera: Geometridae). In *Insect Populations*, ed. J. P. Dempster & I. F. G. McLean, pp. 309–321. Dordrecht: Kluwer Academic Publishers.

Rolley, F., Hodek, I. & Iperti, G. (1974) Influence de la nourriture aphidienne (selon l'âge de la plante-hôte à partir de laquelle les pucerons se multiplient) sur l'induction de la dormance chez *Semiadalia undecimnotata* Schn. (Coleop., Coccinellidae). *Annales de Zoologie et Écologie animale* **6**, 53–60.

Root, R. B. (1967) The niche exploitation pattern of the blue-gray gnatcatcher. *Ecological Monographs* **37**, 317–350.

Rosenheim, J. A., Kaya, H. K., Ehler, L. E., Marois, J. J. & Jaffee, B. A. (1995) Intraguild predation among biological-control agents: theory and evidence. *Biological Control* **5**, 303–335.

Rosenheim, J. A., Limburg, D. D. & Colfer, R. G. (1999) Impact of generalist predators on a biological control agent, *Chrysoperla carnea*: direct observations. *Ecological Applications* **9**,

Rosenheim, J. A., Wilhoit, L. R. & Armer, C. A. (1993) Influence of intraguild predation among generalist insect predators on the suppression of an herbivore population. *Oecologia* **96**, 439–449.

Rothschild, M. (1961) Defensive odours and Müllerian mimicry among insects. *Transactions of the Royal Entomological Society of London* **113**, 101–122.

Rowlands, M. L. J. & Chapin, J. W. (1978) Prey search behavior in adults of *Hippodamia convergens* (Coleoptera : Coccinellidae). *Journal of the Entomological Society* **13**, 309–315.

Růžička, J. (1992) The immature stages of central European species of *Nicrophorus* (Coleoptera, Silphidae). *Acta Entomologia Bohemoslov* **89**,113–135.

Růžička, Z. (1996) Oviposition-deterring pheromone in chrysopids: Intra – and interspecific effects. *European Journal of Entomology* **93**, 161–166.

Růžička, Z. (1997a) Persistence of the oviposition–deterring pheromone in *Chrysopa oculata* (Neur: Chrysopidae). *Entomophaga* **42**, 109–114.

Růžička, Z. (1997b) Recognition of oviposition–deterring allomones by aphidophagous predators (Neuroptera: Chrysopidae, Colleoptera: Coccinellidae). *European Journal of Entomology* **94**, 431–434.

Růžička, Z. & Havelka, J. (1998) Effects of oviposition–deterring pheromone and allomones on *Aphidoletes aphidomyza* (Diptera: Cecidomyiidae). *European Journal of Entomology* **95**, 211–216.

Růžička, Z., Iperti, G. & Hodek, I. (1981) Reproductive rate and longevity in *Semiadalia undecimnotata* and *Coccinella septempunctata* (Coccinellidae, Col.) *Vestník Ceskoslovenske Spolecnosti Zoologicke* **45**, 115–128.

Saâdaoui, H. (1996) Analyse de la ponte et conséquences du cannibalisme des oeufs de la coccinelle *Adalia bipunctata* (L.) Diplôme, Faculté des Sciences agronomiques, Gembloux, Belgium.

Sabelis, M. W. (1981) Biological control of two-spotted mites using phytoseiid predators 1. Modelling the predator–prey interactions at the individual level. PhD thesis, in Agricultural Research Report, Pudoc, Wageningen.

Sabelis, M. W. (1992). Predatory arthropods. In *Natural Enemies. The Population Biology of Predators, Parasites and Diseases*, ed. M. J. Crawley, pp. 225–264. Oxford: Blackwell Scientific Publications.

Saha, J. L. (1987) Studies on the fecundity, hatchability, mortality and longevity of *Menochilus sexmaculatus* Fabr. (Coleoptera: Coccinellidae). *Journal of Aphidology* **1**, 47–50.

Sailer, R. L. (1961) Possibilities for genetic improvements of beneficial insects. In *Germ plasma resources*, American Association for the Advancement of Science, Washington DC, p. 295.

Sallée, B. & Chazeau, J. (1985) Cycle de développement, table de vie, et taux intrinsèque d'accroissement en conditions contrôlées de *Coelophora mulsanti* (Montrouzier), Coccinellidae aphidiphage de Nouvelle-Calédonie (Coleoptera). *Annales de la Société entomologique de France (N.S.)* **21**, 407–412.

Salt, G.(1920) A contribution to the ethology of the Meliponinae. *Transactions of the Entomological Society of London* **77**, 431–470.

Samways, M. J. (1984) Biological and economic value of the scale predator *Chilocorus nigritus* (F.) (Coccinellidae). *Biocontrol News and Information* **5**, 91–105.

Samways, M. J. (1988) A pictorial model of the impact of natural enemies on the population growth rate of the scale insect *Aonidiella aurantii*. *South African Journal of Science* **84**, 270–272.

Samways, M. J. (1989) Climate diagrams and biological control: an example from areography of the ladybird *Chilocorus nigritus* (Fabricius, 1798) (Insecta, Coleoptera, Coccinellidae). *Journal of Biogeography* **16**, 345–351.

Samways, M. J. (1994) *Insect Conservation Biology*. London: Chapman & Hall.

Samways, M. J., Osborn, R. & Saunders, T. L. (1997) Mandible form relative to the main food type in ladybirds (Coleoptera: Coccinellidae). *Biocontrol Science and Technology* **7**, 275–286.

Samways, M. J. & Wilson, S. J. (1988) Aspects of the feeding behaviour of *Chilocorus nigritus* (F.) (Col., Coccinellidae) relative to its effectiveness as a biocontrol agent. *Journal of Applied Entomology* **106**, 177–182.

Sanborn, C. E. (1906) The melon louse and other aphids. *Texas Agricultural Experimental Station Bulletin* **89**.

Sasaji, H. (1967*a*). A revision of the Formosan Coccinellidae (I). The subfamily Sticholotinae with an establishment of a new Tribe (Coleoptera). *Etizenia* No. **25**.

Sasaji, H. (1967*b*). On the Coccinellidae attacking the aphids in Japan and the Ryukyus. *Mushi* **40**, 147–175.

Sasaji, H. (1968*a*) Descriptions of the coccinellid larvae of Japan and the Ryukyus (Coleoptera). *Memoirs of the Faculty of Education of Fukui University, Series 11, Natural History* **18**, 93–135.

Sasaji, H. (1968*b*). A revision of the Formosan Coccinellidae (II). Tribes Stethorini, Aspidimerini and Chilocorini (Coleoptera). *Etizenia* No. **32**.

Sasaji, H. (1971) *Fauna Japonica – Coccinellidae (Insecta: Coleoptera)*. Tokyo: Academic Press of Japan.

Sasaji, H. & McClure, M. S. (1997) Description and distribution of *Pseudoscymnus tsugae* sp. nov. (Coleoptera: Coccinellidae), an important predator of hemlock woolly adelgid in Japan. *Annals of the Entomological Society of America* **90**, 563–568.

Schanderl, H., Ferran, A. & Garcia, V. (1988) L'Élevage de deux coccinelles *Harmonia axyridis* Pallas et *Semiadalia undecimpunctata* Schn. à l'aide d'oeufs d'*Anagasta kuehniella* Zell. tués aux rayons ultraviolets. *Entomologia Experimentalis et Applicata* **49**, 235–244.

Schanderl, H., Ferran, A. & Larroque, M. M. (1985) Les besoins trophiques et thermiques des larves de la coccinelle *Harmonia axyridis* Pallas. *Agronomie* **5**, 417–421.

Schilder, F. A. & Schilder, M. (1928). Die Nahrung der Coccinelliden und ihre Beziehung zur Verwandtschaft der Arten. *Arbeiten aus der biologischen Reichsanstalt für Land- und Forstwirtschaft* **2**, 213–282.

Schröder, F. C., Farmer, J. J., Attygalle, A. B., Smedley, S. R., Eisner, T. & Meinwald, J. (1998) Combinatorial chemistry in insects: a library of defensive macrocyclic polyamines. *Science* **281**, 428–431.

Semyanov, V. P. (1996) Methods of rearing and long storage of a tropical coccinellid *Leis dimidiata* (Fabr.) (Coleoptera, Coccinellidae). *Entomological Review* **75**, 714–720.

Sengonca, C. & Frings, B. (1985) Interference and competitive behaviour of the aphid predators *Chrysoperla carnea* and *Coccinella septempunctata* in the laboratory. *Entomophaga* **30**, 245–251.

Sengonca, C., Kotikal, Y. K. & Schade, M. (1995) Olfactory reactions of *Cryptolaemus montrouzieri* Mulsant (Col., Coccinellidae) and *Chrysoperla carnea* Stephens (Neur., Chrysopidae) in relation to period of starvation. *Anzieger Schädlingskunde Pflanzenschutz, Umweltschutz* **68**, 9–12.

Sengonca, C. & Liu, B. (1994) Responses of the different instar predator, *Coccinella septempunctata* L. (Coleoptera: Coccinellidae), to the kairomones produced by the prey and non-prey insects as well as the predator itself. *Zeitschrift für Pflanzenkrankheiten und Pflanzenschutz* **101**, 173–177.

Sequeira, R. & Dixon, A. F. G. (1997) The population dynamics of tree-dwelling aphids: the importance of seasonality and time scale. *Ecology* **78**, 2603–2610.

Sharma, D. C., Rawat, U. S. & Pawar, A. D. (1990) Effect of temperature and humidity on the development, longevity and predatory potential of *Pharoscymnus flexibilis* Muls. on San José scale. *Journal of Biological Control* **4**, 11–14.

Sherratt, T. N. & Harvey, I. F. (1993) Frequency-dependent food selection by arthropods: a review. *Biological Journal of the Linnean Society* **48**, 167–186.

Sibly, R. M. & Calow, P. (1986) *Physiological Ecology of Animals*. Oxford: Blackwell Scientific Publications.

Simanton, F. L. (1916) *Hyperaspis binotata*, a predatory enemy of the terrapin scale. *Journal of Agricultural Research* **6**, 197–203.

Simberloff, D. & Stiling, P. (1996) How risky is biological control? *Ecology* **77**, 1965–1974.

Simberloff, D. & Stiling, P. (1998) How risky is biological control? Reply. *Ecology* **79**, 1834–1836.

Simpson, R. G. & Burkhardt, C. C. (1960) Biology and evaluation of certain predators of *Therioaphis maculata* (Buckton). *Journal of Economic Entomology* **53**, 89–94.

Singh, L. S., Singh, K. C. & Singh, T. K. (1993). Biology and feeding potential of *Oenopia quadripunctata* Kapur, a coccinellid predator of an oak aphid *Tuberculatus (Acanthocallis) nervatus* Chakrabarti and Raychaudhuri. *Journal of Advanced Zoology* **14**, 7–11.

Skirvin, D. J., Perry, J. N. & Harrington, R. (1997) The effect of climate change on an aphid-coccinellid interaction. *Global Change Biology* **3**, 1–11.

Slobodkin, L. B. (1986) The role of minimalism in art and science. *American Naturalist* **127**, 257–265.

Smith, B. C. (1966a) Significance of variation in weight, size and sex ratio of coccinellid adults. In *Ecology of Aphidophagous Insects*, ed. I. Hodek, pp. 249–251. The Hague: Dr W. Junk.

Smith, B.C. (1966b) Effect of food on some aphidophagous Coccinellidae. In *Ecology of Aphidophagous Insects*, ed. I. Hodek, pp. 75–81. Prague: Academia.

Smith, H. D. & Maltby, H. L. (1964) Biological control of the citrus blackfly in Mexico. *United States Department of Agriculture Technical Bulletin* no. **1311**.

Smith, H. S. (1919) On some phases of insect control by the biological method. *Journal of Economic Entomology* **12**, 288–292.

Smith, H. S. (1939) Insect populations in relation to biological control. *Ecological Monographs* **9**, 311–320.

Smith, H. S. & Armitage, H. M. (1920) Biological control of mealybugs in California. *Monthly Bulletin of the California State Department of Agriculture* **9**(4), 104–161.

Smith, R. F. & Reynolds, H. T. (1972) Effects of manipulation of cotton agro-ecosystems on insect populations. In *The Careless Technology: Ecology and International Development*, ed. M. T. Farvar & J. P. Milton. New York: Natural History Press.

Stadler, B. (1991) Predation success of *Coccinella septempunctata* when attacking different *Uroleucon* species. In *Behaviour and Impact of Aphidophaga*, ed. L. Polgár, R. J. Chambers, A. F. G. Dixon & I. Hodek, pp 265–271. The Hague: SPB Academic Publishers.

Staines, C. L., Rothschild, M. J. & Trumbule, R. B. (1990) A survey of the Coccinellidae (Coleoptera) associated with nursery stock in Maryland. *Proceedings of the Entomological Society of Washington* **92**, 310–313.

Stanley, S. M. (1973). An explanation for Cope's rule. *Evolution* **27**, 1–26.

Stary, P. (1970) *Biology of Aphid Parasites*. The Hague: Dr W. Junk.

Stäubli Dreyer, B., Baumgaertner, J. U., Neuenschwander, P. & Dorn, S. (1997a) The functional responses of two *Hyperaspis notata* strains to their prey, the cassava mealybug *Phenacoccus manihoti*. *Mitteilungen der schweizerischen entomologischen Gesellschaft* **70**, 21–28.

Stäubli Dreyer, B., Neuenschwander, P., Bouyjou, B., Baumgaertner, J. U. & Dorn, S. (1997b) The influence of temperature on the life table of *Hyperaspis notata*. *Entomologia Experimentalis et Applicata* **84**, 85–92.

Stearns, S. C. (1982) *The Evolution of Life Histories*. Oxford: Oxford University Press.

Stearns, S. C. & Koella, J. C. (1986) The evolution of phenotypic plasticity in life-history traits: predictions of reaction norms for age and size at maturity. *Evolution* **40**, 893–913.

van Steenis, M. J. (1992) Biological control of the cotton aphid, *Aphis gossypii* Glover (Hom., Aphididae): pre-introduction evaluation of natural enemies. *Journal of Applied Entomology* **114**, 362–380.

Stephens, D. W. & Krebs, J. R. (1986) *Foraging Theory*. Princeton, NJ: Princeton University Press.

Stern, V. M., Smith, R. F., van den Bosch, R. & Hagen, K. S. (1959) The integration of chemical and biological control of the spotted alfalfa aphids, Part 1. The integrated control concept. *Hilgardia* **29**, 81–101.

Stevens, L. (1992) Cannibalism in beetles. In *Cannibalism, Ecology and Evolution among Diverse Taxa*, ed. M. A. Elgar & B. J. Crespi, pp. 156–175. Oxford: Oxford University press.

Stewart, L. A. & Dixon, A. F. G. (1989) Why big species of ladybird beetles are not melanic. *Functional Ecology* **3**, 165–177.

Stewart, L. A., Dixon, A. F. G., Růžička, Z. & Iperti, G. (1991a) Clutch and egg size in ladybird beetles. *Entomophaga* **36**, 93–97.

Stewart, L. A., Hemptinne, J.-L. & Dixon, A. F. G. (1991b) Reproductive tactics of ladybird beetles: relationships between egg size, ovariole number and developmental time. *Functional Ecology* **5**, 380–385.

Storch, R. H. (1976) Prey detection by fourth stage *Coccinella transversoguttata* larvae (Coleoptera: Coccinellidae). *Animal Behaviour* **24**, 690–693.

Strand, M. R. & Obrycki, J. J. (1996) Host specificity of insect parasitoids and predators. *Bioscience* **46**, 422–429.

Stubbs, M. (1980) Another look at prey detection by coccinellids. *Ecological Entomology* **5**, 179–182.

Subramanian, V. K. (1953) Control of the fluted scale in penninsular India. *Indian Journal of Entomology* **17**, 103–120.

Subramanyam, T. V. (1925) *Coptosoma ostensum*, Dist. and its enemy *Synia melanaria*, Muls. *Journal of the Bombay Natural History Society* **30**, 924–925.

Sundby, R. A. (1968) Some factors influencing the reproduction and longevity of *Coccinella septempunctata* Linnaeus [Coleoptera : Coccinellidae]. *Entomophaga* **13**, 197–202.

Suter, H. & Keller, S. (1990) *Blattläuse und Blattlausfeinde*. Berne: Budenberg.

Tadmor, U., Applebaum, S. W. & Kafir, R. (1971) A gas-chromatographic micro-method for respiration studies on insects. *Journal of Experimental Biology* **54**, 437–441.

Takahashi, K. (1987) Cannibalism by the larvae of *Coccinella septempunctata brucki* Mulsant (Coleoptera: Coccinellidae) in mass-rearing experiments. *Japanese Journal of Applied Entomology and Zoology* **31**, 201–205.

Takahashi, K. (1989) Intra- and interspecific predations of lady beetles in spring alfalfa fields. *Japanese Journal of Entomology* **57**, 199–203.

Tan, C.-C. (1933–4) Notes on the biology of the lady-bird beetle, *Ptychantis axyridis* Pall. *Peking Natural History Bulletin* **8**, 9–19.

Tan, C.-C. & Li, J.-C. (1932–3) Variations in the color patterns in the lady-bird beetles *Ptychanatis axyridis* Pall. *Peking Natural History Bulletin* **7**, 175–193.

Tanigoshi, L. K. & McMurtry, J. A. (1977) The dynamics of predation of *Stethorus picipes* (Coleoptera: Coccinellidae) and *Typhlodromus floridanus* on the prey *Oligonychus punicae* (Acarina: Phytoseiidae, Tetranychidae). *Hilgardia* **45**, 237–261.

Tartar, M., Carey, J. R. & Vaupel, J. W. (1993) Long-term cost of reproduction with and without accelerated senescence in *Callosobruchus maculatus*: analysis of age specific mortality. *Evolution* **47**, 1302–1312.

Tauber, M. J., Tauber, C. A. & Masuki, S. (1986) *Seasonal Adaptation of Insects*. Oxford: Oxford University Press.

Tawfik, M. F. S. (1962). Studies on *Scymnus (Pullus) syriacus* Mars. *Bulletin of the Society of Entomology of Egypt* **46**, 485–504.

Tawfik, M. F. S. & Nasr, S. A. (1973) The biology of *Scymnus interruptus* Goeze (Coleoptera : Coccinellidae). *Bulletin of the Society of Entomology of Egypt* **57**, 9–26.

Taylor, A. J., Müller, C. B. & Godfray, H. C. J. (1998) Effect of aphid predators on oviposition behavior of aphid parasitoids. *Journal of Insect Behavior* **11**, 297–302.

Taylor, R. J. (1984) *Predation*. New York: Chapman & Hall.

Taylor, T. H. C. (1935) The campaign against *Aspidiotus destructor*, Sign., in Fiji. *Bulletin of Entomological Research* **26**, 1–102.

Thompson, W. R. (1951) The specificity of host relations in predaceous insects. *Canadian Entomologist* **83**, 262–269.

Thorpe, W. H. (1930) The biology, post-embryonic development, and economic importance of *Cryptochaetum iceryae* (Diptera, Agromyzidae) parasitic on *Icerya purchasi* (Coccidae, Monophlebini). *Proceedings of the Zoological Society of London* **60**, 929–971.

Toccafondi, P., Covassi, M. & Pennacchio, F. (1991) Studi sugli entomofagi predatori di cocciniglie del gen. *Matsucoccus* Cock. in Italia II. Note bio-etologiche su *Rhyzobius chrysomeloides* (Herbst.) in pinete della Liguria (Coleoptera, Coccinellidae). *Redia* **74**, 599–620.

Tourniaire, R., Ferran, A., Gambier, J., Giuge, L. & Bouftault, F. (2000a) Locomotory behaviour of flightless *Harmonia axyridis* Pallas (Col., Coccinellidae).

Tourniaire, R., Ferran, A., Giuge, L. & Gambier, L. (2000b) A natural flightless mutation in *Harmonia axyridis* Pallas (Col., Coccinellidae).

Townsend, C. R. & Hughes, R. N. (1981) Maximizing net energy returns from foraging. In *Physiological Ecology, an Evolutionary Approach to Resource Use*, ed. C. R. Townsend & P. Calow, pp. 86–108. Oxford: Blackwell Scientific Publications.

Tranfaglia, A. & Viggiani, G. (1973) Dati biologi sullo *Scymnus includens* Kirsch (Coleoptera : Coccinellidae). *Bolletino del Laboratorio di Entomologia Agraria 'Filippo Silvestri' di Portici*, **30**, 4–18.

Trouve, C., Ledee, S., Ferran, A. & Brun, J. (1997) Biological control of the damson-hop aphid, *Phorodon humuli* (Hom.: Aphididae), using the ladybeetle *Harmonia axyridis* (Col. : Coccinellidae). *Entomophaga* **42**, 57–62.

Turnbull, A. L. & Chant, D. A. (1961) The practice and theory of biological control of insects in Canada. *Canadian Journal of Zoology* **39**, 697–753.

Tursch, B., Braekman, J. C. & Daloze, D. (1976) Arthropod alkaloids. *Experientia* **32**, 401–407.

Ueno, H. (1994) Intraspecific variation of P2 value in a coccinellid beetle, *Harmonia axyridis*. *Journal of Ethology* **12**, 169–174.

Ueno, H., Sato, Y. & Tsuchida, K. (1998) Colour-associated mating success in a polymorphic ladybird beetle, *Harmonia axyridis*. *Functional Ecology* **12**, 757–761.

Umeh, E.-D., N. N. (1982) Biological studies on *Hyperaspis marmottani* Fairm. (Col., Coccinellidae), a predator of the cassava mealybug *Phenacoccus manihoti* Mat-Ferr. (Hom., Pseudococcidae). *Zeitschrift für angewandte Entomologie* **94**, 530–532.

Vandenberg, N. J. (1992) Revision of the New World lady beetles of the genus *Olla* and description of a new allied genus (Coleoptera: Coccinellidae). *Annals of the Entomological Society of America* **85**, 370–392.

Vanhove, F. (1998) Impact de la défense des œufs sur la structure des communautés de Coccinellidae. Mémoire d'Ingénieur Agronome (Eaux & Forêts), Faculté universitaire des Sciences agronomiques, Gembloux, Belgium.

Varma, G. C., Vyas, R. S. & Brar, K. S. (1993) Biology of *Menochilus sexmaculatus* (Fabricius) (Coccinellidae : Coleoptera). *Journal of Research Punjab Agricultural University* **30**, 27–31.

Veeravel, R. & Baskaran, P. (1996) Temperature-dependent development, adult longevity, fecundity and feeding potential of two coccinellid predators under laboratory conditions. *Entomon* **21**, 13–18.

Vesey-Fitzgerald, D. (1940) The control of Coccidae on coconuts in Seychelles. *Bulletin of Entomological Research* **31**, 253–283.

Völkl, W. (1995) Behavioural and morphological adaptations of the coccinellid *Platynaspis luteorubra* for exploiting ant-attended resources (Coleoptera: Coccinellidae). *Journal of Insect Behavior* **8**, 653–670.

Völkl, W. & Vohland, K. (1996) Wax covers in larvae of two *Scymnus* species: do they enhance coccinellid larval survival? *Oecologia* **107**, 498–503.

Waage, J. K. & Mills, N. J. (1992) Biological control. In *Natural Enemies: The Population Biology of Predators, Parasites and Diseases*, ed. M. J. Crawley, pp. 412–430. Oxford: Blackwell Scientific Publications.

Watt, J. C. (1986) Beetles (Coleoptera) of the offshore islands of northern New Zealand. In *The Offshore Islands of Northern New Zealand*, ed. A. E. Wright & R.E. Beever, pp. 221–228. Wellington: Department of Lands and Survey.

Werner, E. & Gilliam, J. (1984) The ontogenetic niche and species interactions in size-structured populations. *Annual Review of Ecology and Systematics* **26**, 619–633.

Wetzel, T., Ghanim, A. E.-B. & Freier, B. (1982) Zur Nahrungsaufnahme von *Coccinella septempunctata* L. bei optimlem Angebot von Aphiden der Art *Macrosiphum avenae* Fabr. *Archiv für Phytopathologie und Pflanzenschutz* 18, 89–96.

Wheeler, W. M. (1911) An ant-nest coccinellid (*Brachyacantha quadripunctata* Mels.). *Journal of the New York Entomological Society* 19, 169–174.

Wicklund, C. & Karlsson, B. (1988) Sexual dimorphism in relation to fecundity in some Swedish satyrid butterflies. *American Naturalist* 131, 132–138.

Wille, J. (1926) *Curius (Orcus) zonatus* (Muls.) (Coccinellidae), ein Feind der Schildläusen an Orangenbäumen. *Zeitschrift für angewandte Entomologie*, 12, 357–375.

Williams, G. C. (1966) *Adaptation and Natural Selection*. Princeton, NJ: Princeton University Press.

Wipperfürth, T., Hagen, K. S. & Mittler, T. E. (1987) Egg production by the coccinellid *Hippodamia convergens* fed on two morphs of the green peach aphid, *Myzus persicae*. *Entomologia Experimentalis et Applicata* 44, 195–198.

Witte, L., Ehmke, A. & Hartmann, T. (1990) Interspecific flow of pyrrolizidine alkaloids. *Naturwissenschaften* 77, 540–543.

Wratten, S. D. (1973) The effectiveness of the coccinellid beetle, *Adalia bipunctata* (L.), as a predator of the lime aphid, *Eucallipterus tiliae* (L). *Journal of Animal Ecology* 42, 785–802.

Wratten, S. D. (1976) Searching by *Adalia bipunctata* (L) (Coleoptera: Coccinellidae) and escape behaviour of its aphid and cicadellid prey on lime (*Tilia × vulgaris* Hayne). *Ecological Entomology* 1, 139–142.

Wright, E. J. & Laing, J. E. (1978) The effects of temperature on development, adult longevity and fecundity of *Coleomegilla maculata lengi* and its parasite *Perilitus coccinellae*. *Proceedings of the Entomological Society of Ontario* 109, 33–47.

Wright, E. J. & Laing, J. E. (1982) Stage-specific mortality of *Coleomegilla maculata lengi* Timberlake on corn in Southern Ontario. *Environmental Entomology* 11, 32–37.

Wyss, E. (1995) The effects of weed strips on aphids and aphidophagous predators in an apple orchard. *Entomologia Experimentalis et Applicata* 75, 43–49.

Wyss, E., Villiger, M., Hemptinne, J.-L. & Müller-Schärer, H. (1999) Effects of augmentative releases of eggs and larvae of the two-spot ladybird beetle, *Adalia bipunctata*, on the abundance of the rosy apple aphid, *Dysaphis plantaginea*, in organic apple orchards. *Entomologia Experimentalis et Applicata* 90, 167–173.

Yasuda, H. & Dixon, A. F. G. (2000) Sexual size dimorphism in ladybirds. *Oikos*.

Yasuda, H. & Ohnuma, N. (1999) Effect of cannibalism and predation on the larval performance of two ladybirds. *Entomologia Experimentalis et Applicata* 93, 63–67.

Yasuda, H. & Shinya, K. (1997) Cannibalism and interspecific predation in two predatory ladybirds in relation to prey abundance in the field. *Entomophaga* 42, 155–165.

Yinon, U. (1969) Food consumption of the armoured scale lady-beetle *Chilocorus bipustulatus* (Coccinellidae). *Entomologia Experimentalis et Applicata* 12, 139–146.

Yoshida, H. A. & Mau, R. F. L. (1985) Life history and feeding behavior of *Nephaspis amnicola* Wingo. *Proceedings of the Hawaiian Entomological Society* 25, 155–160.

Zhao, D.-X. & Wong, Z.-W. (1987) Influence of temperature on the development of the coccinellid beetle, *Scymnus hoffmanni* Weise. *Acta Entomologica Sinica* 30, 47–54.

Zhou, X., Honěk, A., Powell, W. & Carter, N. (1995) Variations in body length, weight, fat content and survival in *Coccinella septempunctata* at different hibernation sites. *Entomologia Experimentalis et Applicata* 75, 99–107.

Taxonomic index

Predators

Adalia 30
Adalia bipunctata 14,17, 19, 23, 29, 30,
 40, 41, 42, 43, 47, 48, 50, 51, 52, 54,
 56, 73, 75, 79, 84, 90, 91, 95, 96, 97,
 98, 101, 105, 106, 107, 120, 123, 126,
 133, 134, 140, 141, 142, 146, 149, 167,
 174, 176, 183, 186, 187, 188, 189, 196,
 206, 212, 213
Adalia decempunctata 17, 86, 87, 98, 122,
 130, 159, 176, 183, 184
Adonis variegata 85, 98
Ailocaria 58
Ailocaria hexaspilota 148
Anatis ocellata 100, 176
Anisosticta novemdecimpunctata 32

Calvia decemguttata 84, 185
Calvia quatuordecimguttata 68, 84, 98,
 185
Cheilomenes lunata 54
Chilocorus bipustulatus 54, 73, 95
Chilocorus cacti 73
Chilocorus distigma 73
Chilocorus infernalis 73
Chilocorus nigritus 13, 73, 95, 97, 99, 104,
 194
Cleidostethus meliponae 7
Cleothora notata 48
Clitostethus arcuatus 15, 54
Coccinella 30
Coccinella californica 54
Coccinella leonina 194
Coccinella maculata 96
Coccinella magnifica 30, 69, 93, 105, 105
Coccinella novemnotata 194
Coccinella quinquepunctata 98
Coccinella septempunctata 10, 13, 21, 22,
 28, 29, 30, 31, 34, 39, 40, 48, 54, 69,
 73, 75, 76, 85, 95, 96, 97, 98, 99, 100,
 101, 102, 105, 106, 110, 111, 112, 113,
 115, 119, 121, 122, 123, 124, 125, 126,
 127, 187, 189, 193, 194, 195, 196, 206,
 211, 213
Coccinella septempunctata brucki 14, 48,
 180, 181, 183
Coccinella transversalis 54, 185, 186
Coccinella transversoguttata 96, 196
Coccinella trifasciata 3
Coccinella undecimpunctata 69, 183, 184,
 191, 194
Coelophora inaequalis 69
Coelophora mulsanti 19, 69
Coelophora quadrivittata 48, 54, 69, 202
Coleomegilla maculata 123, 163, 213
Coleomegilla maculata lengi 148, 177
Cryptolaemus montrouzieri 69, 73, 99,
 100, 104, 211
Cycloneda sanguinea 54, 123, 178

Delphastus pusillus 123
Diomus 100
Diomus hennesseyi 73, 206

Epilachna 15
Epilachna reticulata 13
Exochomus 100
Exochomus flaviventris 73, 206, 211, 212
Exochomus quadripustulatus 30

Harmonia axyridis 31, 41, 48, 52, 54, 73,
 95, 96, 97, 100, 110, 114, 131, 135,
 142, 143, 145, 146, 148, 167, 180, 181,
 182, 183, 196, 206
Harmonia quadripunctata 32
Henosepilachna 15
Hippodamia convergens 27, 28, 29, 41, 80,
 123, 148, 178, 211, 212
Hippodamia quinquesignata 40, 48, 143
Hippodamia tridecimpunctata tibialis 148
Hyperaspis pantherina 197
Hyperaspis trifurcata 33
Hyperaspis raynevali 54, 73, 206

Leis conformis 73
Leis dimidiata 42, 44
Leptothea galbula 15

Megalocaria dillata 33
Menochilus sexmaculatus 15, 20, 23, 29,
    48, 54, 73, 75, 185, 186

Nephaspis aculatus 48
Nephus reunioni 73

Olla v-nigrum 49, 206
Orcus chalybeus 9

Pharoscymnus numidicus 49, 54
Platynaspis 9, 12, 13
Platynaspis luteorubra 93
Propylea japonica 17, 29, 37, 38, 42, 158,
    159, 180, 181
Propylea quatuordecimpunctata 29, 46,
    49, 73, 98, 123, 142, 176, 206, 213
Pseudoscymnus kurohime 33, 40
Pseudoscymnus tsugae 49
Pseudosynonychia 58

Rhizobius 203, 210
Rhizobius lophanthae 54, 196, 206
Rhizobius ventralis 202
Rodolia 178, 197, 203, 210
Rodolia amabilis 194
Rodolia cardinalis 1, 9, 30, 99, 191, 194,
    202
Rodolia iceryae 49.
Rodolia limbata 191, 192

Scymnus 33, 58, 70, 72
Scymnus frontalis 19, 40
Scymnus hoffmani 15, 19
Scymnus interruptus 79
Scymnus marginocollis 54
Semiadalia undecimnotata 27, 28, 149,
    206
Stethorus 12, 58, 72
Stethorus madecassus 49, 54
Stethorus picipes 49, 54
Stethorus punctillum 49, 54
Stethorus punctorum 94
Subcoccinella 15

Tytthaspis 12
Tytthaspis sedecimpunctata 13

Prey

Acythosiphon pisum 3, 95, 123
Aonidiella aurantii 95
Aphis craccivora 101
Aphis fabae 95, 101, 213
Aphis gossypii 180, 181, 211
Aphis jacobaeae 33, 34
Aphis nerii 85
Aphis spiraephaga 101

Carulaspis minima 196
Ceroplastes floridensis 95
Cinara pini 105
Coccus hesperidum 95

Dactylopius 33
Diuraphis noxia 123
Drepanosiphum platanoidis 92, 173, 174
Dysaphis plantaginea 212

Ephestia kuehniella 211
Eucallipterus tiliae 91
Euceraphis punctipennis 91, 105

Hyalopterus pruni 84, 87, 105

Icerya aegyptiaca 191
Icerya purchasi 1, 99, 191, 192, 194

Lepidosaphes newsteadi 196

Macrosiphum albifrons 85
Macrosiphum rosae 211, 212
Megoura viciae 84
Microlophium carnosum 193
Myzus cerasi 105
Myzus persicae 102

Orthesia insignis 196
Ostrinia nubilalis 96

Phorodon humuli 101, 105, 212
Planococcus citri 99
Prociphilus tesselatus 203
Pseudococcus citri 211
Pterocomma salicis 105

Rhopalosiphum maidis 163
Rhopalosiphum padi 105, 193

Saissetia oleae 95
Schizolachnus pineti 105, 114

Subject index

abundance 125–6
ageing 19–21, 77, 80, 88
alkaloids 30, 108, 183, 184
  defence 32
  alimentary canal 14, 120
alkanes 33, 187
ants 105
attack
  coefficient 123
  rate 82, 165, 204

biological control 4, 159, 183, 190, 199, 211
  agents 2, 4, 83, 155, 159, 165, 169, 175, 197–8, 200, 204
  augmentative 191, 211
  classical 190
  conservation 196
  evolutionarily stable 216
  theory 193, 194
birth weight 47
bleeding, reflex 30–1
body size 37, 39, 41, 47, 90, 169
  distribution 55, 58, 65
  growth rate, relative 37, 39, 47

cannibalism 108, 125, 127, 130, 133–9, 141, 146, 165, 187, 213
  avoidance of 142
  cost of 132
  risk of 132, 143
  sibling 132
  survival 135, 137, 149
  theory 130
colour 6, 32
competition 57, 108, 173, 187, 193–4, 199
conversion efficiency 205, 206
defence 29
  alkaloids 30, 108

aposematic 32
  sequestering 33
  trade-off 31
demand rate 204
development 15, 69, 155, 157, 166, 210
  artificial selection for 69
  gonadal 50
  phylogenetic 71
  constraint
  egg weight 27, 39
  rate 37, 70
    fast 41, 67
    isomorphy 15
    slow 41, 67
  survival 17
  temperature, effect of 15, 37, 69, 70
  trade-off 27
dimorphism, sexual 42, 44
  birth weight 47
  fecundity advantage 45, 52
  food consumption 42, 54
  food quality 43
  gonadal constraint 45, 50
  protandry 45, 46
  relative growth rate 52
  temperature 42
  time and energy constraint 46, 52

egg window 106
eggs 8, 15, 108, 140, 142
  adult size 25, 63
  ageing 20
  cluster number 20, 21, 23–4
  duration of incubation 142
  food stress 22, 39
  hatching period 143
  minimum size 21, 39, 40
  ovariole number 23, 24
  protection of 33, 183
  size 21, 22, 26, 27, 39

eggs (*cont.*)
   social feeding 108
encounter rate 119

fat reserves 21, 27, 28, 29, 40, 95
fecundity 18, 20, 207
   adult size 23
   age, effect of 19, 20, 21, 80
   food consumption 20, 21
   longevity 19, 27, 74, 80
   ovariole number 24
   prey quality 23
   trade-off 19, 21
   triangular fecundity function 18, 20
fitness 4, 28, 82, 89, 97, 109, 125, 126,
       132, 135, 147, 151, 213, 216
   inclusive 132, 134, 135
food supply 37, 39, 41, 72, 167
foraging behaviour 82, 95, 115
   abundance 126
   aggregative response 3, 94, 111
   aphid alarm pheromone 92, 101, 108
   area-restricted search 99–101, 109,
      114, 119–120
   arrestants 99, 110, 113
   attack coefficient 123
   capture efficiency 104
   dispersal 82, 94
   encounter rate 115, 119, 120
   energy intake 115, 119, 122
   extensive search 109, 113, 122
   feeding time 121
   functional response 82
   gain, rate of 115
   handling time 113, 118, 121, 157
   harvesting, rate of 94, 115, 119
   hunger 88, 89, 111, 113, 115, 119, 122
   intensive search 109, 120
   larval 109
   location of prey 97, 109
      olfactory 99, 112
      visual 100
   memory window 115, 119
   optimal 104, 112, 120, 122, 124, 147,
      160, 169
   prey
      abundance of 112
      quality of 101
      recognition of 83
   relative risk 84
   speed of movement 109
   survival 124
   switching 93
   systematic search 111
functional response 82, 123, 157, 160,
      162, 164, 166

generation-specific strategies 29
generation time ratio 169, 208

growth rate, relative 37, 39, 47, 52, 73,
   149, 207
guilds
   aphidophagous 97, 108
gut capacity 120

habitat 93, 97, 173
hunger 88–9, 111, 113, 115, 119, 122,
   135, 185

ice-box hypothesis 147
increase, intrinsic rate of 39, 41, 96,
   165, 166, 168, 171, 208
   artificial selection for 41
   developmental rate 168, 208
   temperature, effect of 166
instars, number of 8

legs 14, 185, 202
   length of 68
life cycle 8
life history 8
   parameters 165, 171, 207
   slow–fast continuum 67
life tables 130, 133, 140, 147
longevity 18
   ageing 77, 88
   development 77
   fecundity 19, 55, 74, 77, 78, 80
   food quality 19, 75
   mating success 53
   reproduction 27, 76
   trade-off 19, 76

male-killer disease 142
mating success 52, 53, 56
mimicry
   Batesian 31, 32
   Müllerian 30, 32, 109
morphology 10
   alimentary canal 14
   legs 14, 185, 202
   mouthparts 12
mortality 18
mouthparts 12, 15
movement, speed of 109
myrmecophily 69, 92, 105

Nicholson & Bailey 155

odour 30, 95, 99, 104
ovarioles 25, 39, 52
overwintering 27, 32, 126
   fat reserves 28, 41
   fecundity 76
   gender 28
   generation-specific strategies 29
   hibernation sites 27, 29
   photoperiod 27

reproduction 29
    size 28
    survival 28
oviposition 8, 100, 164
    aphid abundance, peak 107
    artificial diet, effect of 96
    larval tracks 103, 106
oviposition-deterring pheromone 106,
    107, 147

parasitoids 151
patch quality 97, 102
    assessment of 105
    phenology 106
    prey abundance 106
    time-scale 102, 105
    uncertainty of 125
pest management, integrated 214
pheromones
    aphid alarm 92, 101, 108
    oviposition-deterring 106, 107, 147
phylogeny 7
plasticity, intraspecific 36, 41
predation, intraguild 88, 97, 173, 187
    asymmetry 177
    cost 183
    guild structure 173
    predator facilitation 179
    predator–predator interactions 175
predator, top 173, 188
predator–prey interactions 126, 151
    aggregative response 3
    bottlenecks 168, 208
    density 139, 141
    generation time ratios 165, 169
    minimalism 165
    Nicholson & Bailey 155
    parasitoids 151
    prey
        equilibrium 156
        rate of increase 156
    reproductive response 3
    searching efficiency 156
    temperature, effect of 3, 166
    theory 152
prey 6, 14, 89–90, 93
    alternative 89, 93
    essential 89, 93
    nursery 93
    peripheral 89
    quality of 90
    switching 93, 138
    toxic 89
prey recognition 83
    alkanes 84
    odour 99
prey specificity 83, 88, 96, 97, 202
    habitat quality 94

mobility of prey 90
plant structure 90
relative risk 84
size 58, 64
pupae 146

reproduction 18, 156
    interspecific 23
    intraspecific 23
        adult weight 23, 26
        ovariole number 23
    overwintering 29
reproductive rate 26, 52, 55, 74
    biomass 26, 77
    response 3
respiratory rate 64, 73
response
    aggregative 3, 94, 111
    functional 82, 123, 157, 160, 162, 164,
        166
    numerical 91, 155
    olfactory 99, 104
    reproductive 3
    visual 97, 100

satiation 91, 119, 120
search
    area-restricted 99–101, 109, 114,
        119–20
    extensive 109, 113, 122
    intensive 109, 120
    systematic 111
searching efficiency 123, 156, 200
sexual maturity 50
    gonadal development 51
slow–fast continuum 42, 67, 80, 207
    developmental time 69
    fecundity 74
    longevity 74
    metabolic rate 69
    speed of movement 58, 68, 185,
        202
    trade-off 76
social feeding 108
speed of movement 58, 68
structure 6
survival 17, 64, 124, 159
    duration of development 17, 90
    food supply 17, 18
    slow-growth-high-mortality 18

temperature 37, 39, 40
thermal thresholds 3
toxicity of prey 89
transit time 108

visual response 97, 100
voracity 14, 169, 171, 204, 205